Longman
Mathematics
for IGCSE
Practice Book 2

D A Turner, I A Potts

PEARSON

Longman

Pearson Education Limited
Edinburgh Gate
Harlow
Essex
CM20 2JE
England
www.longman.co.uk

First published 2007

ISBN: 978-1-4058-6504-3

Cover design by Juice Creative Ltd.
Typeset by Tech-Set, Gateshead
Printed in Great Britain by Scotprint, Haddington

The publisher's policy is to use paper manufactured from sustainable forests.

We are grateful for permission from Edexcel Limited to reproduce past exam questions. All such questions have a reference in the margin. Edexcel Limited can accept no responsibility whatsoever for the accuracy of any solutions or answers to these questions.

Every effort has been made to trace the copyright holders and we apologise in advance for any unintentional omissions. We would be pleased to insert the appropriate acknowledgement in any subsequent edition of this publication.

Contents

Unit 4

Number 1

Inverse proportion and recurring decimals
Exercise 1

1 It takes one person four days to fully tile a bathroom. To tile a similar bathroom, how long would it take

   a two people   **b** four people   **c** eight people?

2 It takes one person five days to fully weed a garden. To weed a similar garden, how long would it take

   a two people   **b** four people   **c** ten people?

3 The Humber Bridge created about 12 000 man-years of employment in England. In theory it could have been built by 2000 workers in six years. Copy and complete the following table for the bridge's construction.

Number of years, n	Number of men, m
1	
	6000
3	
	3000
6	
	m
n	

4 It takes one person five days to lay a 10 m brick garden path.

   a How long would it take to lay a similar 10 m garden brick path using

      i two people   **ii** four people

      iii five people   **iv** p people?

   b How many people would it take to lay a similar 40 m garden path in

      i two days   **ii** four days

      iii five days   **iv** d days?

5 A helicopter travels between Paris and Rotterdam at a constant speed of 150 km/hr for 3 hours.

   a How long would the helicopter take to travel between Paris and Rotterdam travelling at 50 km/hr?

   b Calculate the speed of the helicopter travelling from Paris to Rotterdam if the journey took 5 hours.

6 Express the following fractions as recurring decimals using dot notation

   a $\frac{1}{3}$ **b** $\frac{1}{9}$

   c $\frac{1}{12}$ **d** $\frac{1}{15}$

7 Change these recurring decimals into fractions.

   a $0.\dot{4}$ **b** $0.\dot{7}$

   c $0.\dot{5}\dot{3}$ **d** $0.\dot{8}\dot{2}$

8 Change these recurring decimals into fractions.

   a $0.\dot{3}0\dot{1}$ **b** $0.\dot{7}0\dot{7}$

   c $0.0\dot{3}\dot{5}$ **d** $0.00\dot{4}0\dot{9}$

Exercise 1★

1 It takes two women four days to build an 8 m dry-stone wall. Use this information to copy and complete the following table.

Number of women, *w*	Length of dry-stone wall, *x* m	Time of construction, *t* days
1	8	
3		6
4	24	
6		2
	32	12
w	*x*	
w		*t*
	x	*t*

2 A single bumble bee travels 150 km to produce 1 g of honey. Use this information to copy and complete the following table for a similar colony of bumble bees.

Number of bees, *b*	Length of bee's journey, *x* km	Mass of honey, *m* g
10	150	
20		50
	750	500
500		1000
10^6	1000	
b	*x*	
b		*m*
	x	*m*

3 A rugby pitch can be cut with two grass-cutters in 30 minutes. Use this information to copy and complete the following table.

Number of grass-cutters, *n*	Number of rugby pitches, *r*	Time, *t* hours
2	1	0.5
	2	1.5
4	4	
	r	*t*
n		*t*
n	*r*	

4   A house in the UK produces on average 20 kg of waste per week. Use this information to copy and complete the following table:

Number of houses, h	Mass of waste, w kg	Time, t weeks
1	20	1
100		10
	10^6	52
h	w	
	w	t
h		t

5   Convert the following recurring decimals into fractions.
   a $0.3\dot{2}\dot{1}$ b $0.7\dot{5}$
   c $0.3\dot{7}\dot{8}$ d $1.0\dot{2}\dot{5}$

6   Express $0.\dot{2}5\dot{4} \div 0.\dot{7}1\dot{3}$ as a fraction.

Algebra 1

Proportion

Exercise 2

1 y is directly proportional to x. If $y = 18$ when $x = 2$, find
       a the formula for y in terms of x **b** y when $x = 10$
       c x when $y = 45$.

2 y varies as the square of x. If $y = 10$ when $x = 5$, find
       a the formula for y in terms of x **b** y when $x = 15$
       c x when $y = 15$.

3 p is directly proportional to the square root of q. If $p = 100$ when $q = 25$, find
       a the formula for p in terms of q **b** p when $q = 64$
       c q when $p = 50$.

4 p is inversely proportional to q. If $p = 5$ when $q = 10$, find
       a the formula for p in terms of q **b** p when $q = 20$
       c q when $p = 20$.

5 a is directly proportional to b. Copy and complete the following table.

b	10	15	
a	200		600

6 The vertical distance a stone falls from a cliff, d m, is directly proportional to the square of the time, t seconds. If the stone falls 45 metres in 3 seconds, find
       a the formula for d in terms of t **b** d when $t = 2$ seconds
       c t when $d = 90$ m.

7 The cost in $\$C$ of a circular wedding cake is directly proportional to its diameter d cm. If a 40 cm diameter cake costs $60, find
       a the formula for C in terms of d
       b the cost of a cake with a diameter of 650 mm
       c the diameter of an $80 cake.

8 The price of an art book, £p varies directly as the square of the number of coloured pictures, n, that it contains. If a £150 art book contains ten coloured pictures, find
       a the formula for p in terms of n
       b the price of an art book containing 12 coloured pictures
       c the number of coloured pictures contained in a £600 art book.

Exercise 2★

1 y is directly proportional to the cube of x. If $y = 32$ when $x = 2$, find
   a the formula for y in terms of x
   b y when $x = 4$
   c x when $y = 864$.

2 y is inversely proportional to the square of x. If $y = 20$ when $x = 2$, find
   a the formula for y in terms of x
   b y when $x = 4$
   c x when $y = 0.5$.

3 p is inversely proportional to the square root of q. If $p = 500$ when $q = 25$, find
   a the formula for p in terms of q
   b p when $q = 100$
   c q when $p = 50$.

4 p squared is inversely proportional to the cube of q. If $p = 10$ when $q = 2$, find
   a the formula for p in terms of q
   b p when $q = 4$
   c q when $p = 5$.

5 a is inversely proportional to the cube root of b. Copy and complete the following table.

b	125	8	
a	2		10

6 The energy of a meteorite, e kilo-joules (kJ) is directly proportional to the square of its speed, v m/s.
   If a meteorite travelling at 10 m/s has energy of 50 kJ, find
   a the formula for e in terms of v
   b e when $v = 50$ km/h
   c v when the energy is 10^6 J

7 The population of a termite hill, n thousand, is inversely proportional to the square of its age, t years.
   If a termite hill has a population of one million after six months, find
   a the formula for n in terms of t
   b n after two years
   c t when the population is one thousand.

8 Tidal waves (tsunamis) are the result of earthquakes in the sea bed. Their speed, v m/s, is directly
   proportional to the square root of the ocean depth, d m. If a tsunami travels at 9.8 m/s at an ocean
   depth of 10 m, find
   a the formula for v in terms of d
   b v when   **i** $d = 50$ m   **ii** $d = 1$ km
   c d when $v = 1$ m/s.
   d the ocean depth at the point where the fastest ever tsunami was recorded at 790 km/h.

Graphs 1

Cubic graphs and reciprocal graphs

Exercise 3

1 Copy and complete the following table for the equation $y = x^3 - 3x^2 + 3$ and use it to draw the graph.

x	-2	-1	0	1	2	3	4
y	-17		3				

2 Copy and complete the following table for the equation $y = 2x^3 - 3x^2 - 4x + 1$ and use it to draw the graph.

x	-2	-1	0	1	2	3	4
y	-19		1				

3 Copy and complete the table for the equation $y = \dfrac{12}{x}$ and use it to draw the graph.

x	-4	-3	-2	-1	0	1	2	3	4
y	-3				$-$				

4 Copy and complete the table for the equation $y = \dfrac{100}{x}$ and use it to draw the graph.

x	1	4	8	12	16	20
y	100					5

5 *Sketch* the following graphs

a $y = x^3$

b $y = \dfrac{1}{x}$

c $y = -x^3$

d $y = -\dfrac{1}{x}$

6 The profit, p ($1000s), made by Giggle, a new internet search engine, t months after its opening is given by $p = \dfrac{20}{t}$, valid for $1 \leqslant t \leqslant 12$.

a Copy and complete the table and use it to draw the graph of p against t.

t	1	2	4	6	8	10	12
p	20		5				

b Use your graph to find

   i after how many months Giggle make a profit of $15 000

   ii Giggle's profit after nine months

c Giggle's profit at one month is reduced by x% after five months. Find x.

7   The depth, d m, of water in a tidal harbour, t hours after midnight is given by $d = t^3 - 8t^2 + 12t + 18$, for $0 \leqslant t \leqslant 6$.

a   Copy and complete the table and use it to draw the graph of d against t.

t	0	1	2	3	4	5	6
d	18		18				

b   Use your graph to find the maximum depth of the harbour water and at what time this occurs.

c   An oil tanker requires at least 10 m water depth to dock in the harbour. Between what times should the oil tanker be kept out of the harbour?

Exercise 3★

1   a   If $y = x^3 + ax^2 + 2x + 9$, where a is a constant, for $-2 \leqslant x \leqslant 5$, use the table to find the value of a and hence copy and complete the table.

x	-2	-1	0	1	2	3	4	5
y			9					19

b   Use the table to draw the graph.

c   Find the values of x at which the curve cuts the x-axis.

2   a   If $y = x^3 + ax^2 + 9x + b$, where a and b are constants, for $-1 \leqslant x \leqslant 6$, use the table to find the values of a and b and hence copy and complete the table.

x	-1	0	1	2	3	4	5	6
y		8						26

b   Use the table to draw the graph.

c   Find the values of x at which the curve cuts the x-axis.

3   a   If $y = \dfrac{k}{x}$, where k is a constant, for $1 \leqslant x \leqslant 6$, use the table to find the value of k and hence copy and complete the table.

x	1	2	3	4	5	6
y						2

b   Use the table to draw the graph.

4   a   If $y = ax + \dfrac{4}{x}$, where a is a constant, for $1 \leqslant x \leqslant 6$, use the table to find the value of a and hence copy and complete the table.

x	1	2	3	4	5	6
y	8					

b   Use the table to draw the graph.

c   Find the co-ordinates of the lowest value of y for this graph.

5 **a** The temperature $T\,(°C)$, in Paris on 1st April t hours after 07:00 is given by the equation
$T = t^3 + at^2 + 8t + b$, for $0 \leqslant t \leqslant 5$, where a and b are constants. Use the table to find the value
of a and b and hence copy and complete the table.

t	0	1	2	3	4	5
T	10					25

b Use the table to draw the graph of T against t.

c Find the lowest temperature in this time period and when it occurs.

6 **a** The speed $V\,(\text{m/s})$ of a cyclist on a mountainous section of the
Tour de France, t minutes after 12:00 is given by the equation
$v = at + \dfrac{b}{t}$, for $1 \leqslant t \leqslant 10$, where a and b are constants.

Use the table to find the values of a and b and hence copy and
complete the table.

t	1	2	4	6	8	10
v	25					52

b Use the table to draw the graph of v against t.

c Find

i the speed at which the cyclist is travelling on the steepest section and the time at which this
occurs

ii the time after which the cyclist's speed is at least 25 m/s.

Shape and space 1

Congruence and circle theorems

Exercise 4

In Questions 1–8 find the angle marked *a*.

1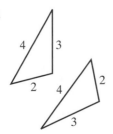

2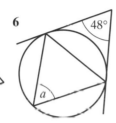

3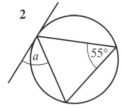

4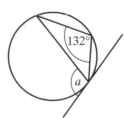

5

6

7

8

In Questions 9–16 find the length marked *x*.

9

10

11

12

13

14

15

16

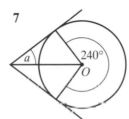
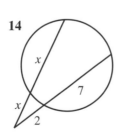
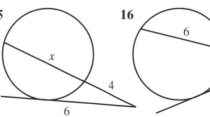

In Questions 16–20 state whether the triangles are congruent or not and give the condition, e.g. SSS, SAS

17 a   **b**

18 a
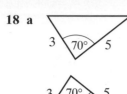

b

19 a
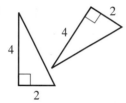

b
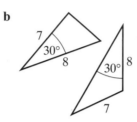

20 a
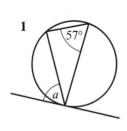

b
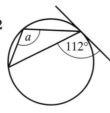
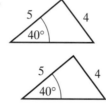

Exercise 4★

In Questions 1–8 find the angle marked a.

1

2

3

4

5

6

7
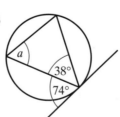

8
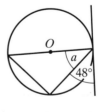

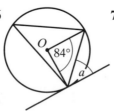
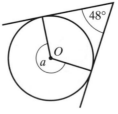
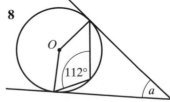

In Questions 9–16 find the length marked x.

9
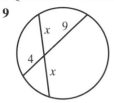

10
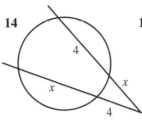

11
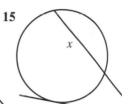

12
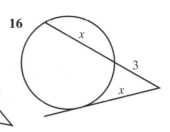

13

14

15

16
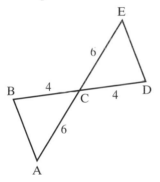

17 In the diagram, BCD and ACE are straight lines. Use congruent triangles to prove that AB is parallel to DE.

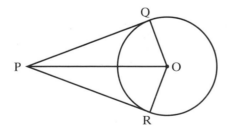

18 In the diagram, ABCD is a paralleloram. Use congruent triangles to prove that AX = XD

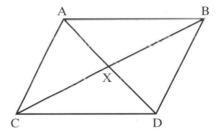

19 In the diagram, PQ and PR are tangents to the circle. Use congruent triangles to prove angle QPO = angle RPO

20 In the diagram, ABCD is a parallelogram. P, Q, R and S are the midpoints of the sides. Use congruent triangles to prove that PQRS is also a parallelogram.

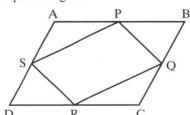

Problems, indentifying sets and set-builder notation

Exercise 5

In Questions 1–4, draw a Venn diagram to represent the information, and use it to solve the problem.

1 In a class of 30 students, six play the drums, twelve play the guitar while four students play both. How many students play neither instrument?

2 In a hockey team of eleven players, nine have brown hair, three have green eyes and one has neither. How many players have both brown hair and green eyes?

3 In a car club of 50 cars there are ten blue cars and fifteen Fords. 28 cars are neither blue nor a Ford. How many blue Ford cars are there?

4 In a group of 40 teenagers all except one have a mobile phone. 18 of the phones can take videos and 12 of the phones are pink. 13 phones are neither pink nor can take videos. How many pink phones can't take videos?

5 On separate copies of the diagram, shade the following sets.

    a A ∩ B   **b** A′ ∩ B

    c A ∩ B′   **d** A′ ∩ B′

    e (A ∩ B)′

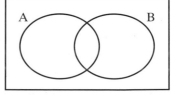

6 On separate copies of the diagram, shade the following sets.

    a A ∪ B   **b** A′ ∪ B

    c A ∪ B′   **d** A′ ∪ B′

    e (A ∪ B)′

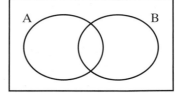

7 On separate copies of the diagram, shade the following sets.

    a A ∩ B   **b** A′ ∩ B

    c A ∩ B′   **d** A′ ∩ B′

    e (A ∩ B)′

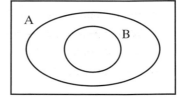

8 On separate copies of the diagram, shade the following sets.

    a A ∪ B   **b** A′ ∪ B

    c A ∪ B′   **d** A′ ∪ B′

    e (A ∪ B)′

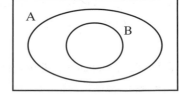

9 List the following sets where $\xi = \{\text{Integers}\}$

  a $A = \{x: -1 \leqslant x < 2\}$ **b** $B = \{x: 4 < 2x \leqslant 10\}$

  c $C = \{x: -5 < 3x \leqslant 3\}$ **d** $D = \{x: 0 < x^2 \leqslant 20\}$

10 Express the following sets in set builder notation where $\xi = \{\text{Integers}\}$

  a $A = \{\text{integers less than 5}\}$ **b** $B = \{\text{integers greater or equal to } -8\}$

  c $C = \{\text{integers between } -2 \text{ and } 4\}$ **d** $D = \{\text{integers between 3 and 8 inclusive}\}$

Exercise 5★

In Questions 1–4, draw a Venn diagram to represent the information, and use it to solve the problem.

1 A group are on an activity holiday. Eighteen of the group do horse riding while eight do quad biking but not horse riding. Five of the group do neither of these activities. How many are in the group?

2 There are 50 children in a youth club. 19 have blue eyes, seven have green eyes and six have red hair. Five of the group have green eyes but not red hair and 23 of the group have neither blue eyes, green eyes nor red hair. x have blue eyes and red hair.

  a Write down an expression, in terms of x, for the number who have red hair but neither blue eyes nor green eyes.

  b Find x, the number with blue eyes and red hair.

3 40 teenagers are going on a coach outing, and all of them like at least one out of peppermints, chocolates and toffees. All those who like chocolates also like peppermints, but nobody likes chocolates and toffees. 12 like toffees, 33 like peppermints and ten like chocolates. x like peppermints and toffees.

  a Write down an expression, in terms of x, for the number who like toffees only.

  b Find the value of x.

4 Mr Weinbuff has 60 bottles of wine in his wine rack. He particularly likes wine containing the grape varieties Shiraz, Merlot and Pinot Noir, though 15 of the bottles contain none of these varieties. Ten bottles contain Shiraz only, seven contain Merlot only and twelve contain Pinot Noir only. Four bottles contain Shiraz and Pinot Noir but not Merlot, six bottles contain Shiraz and Merlot and nine bottles contain Merlot and Pinot Noir. x bottles contain all three grape varieties.

  a Write down an expression, in terms of x, for the number of bottles containing Shiraz and Merlot but not Pinot Noir.

  b Find the value of x.

5 Two sets A and B are such that $A \cap B \neq \varnothing$. By shading Venn diagrams, prove that

  a $A' \cap B' = (A \cup B)'$ **b** $(A \cup B')' = A' \cap B$

6 Do the conclusions of question **5** hold when

  a $B \subset A$ **b** $A \cap B = \varnothing$?

7 On separate copies of the diagram, shade the following sets

  a $(A \cup C) \cap B$ **b** $(A \cup B)' \cap C$

  c $(A \cap B)' \cap C$ **d** $(A \cap C)' \cup B$

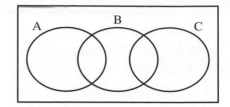

8 On separate copies of the diagram, shade the following sets

  a $(A \cup B)' \cap C$ **b** $(A \cup C)' \cap B$

  c $(A \cap C) \cup B$ **d** $(A \cap B)' \cup C$

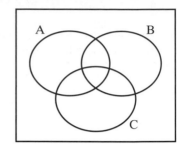

9 List the following sets where $\xi = \{\text{Integers}\}$

  a $A = \{x: -10 \leqslant 5x < 20\}$ **b** $B = \{x: x = 2y \text{ and } 4 < y \leqslant 7\}$

  c $C = \{x: x^2 - 4 = 0\}$ **d** $D = \{x: -1 < \frac{1}{2}x \leqslant 3\}$

10 Express the following sets in set builder notation where $\xi = \{\text{Integers}\}$

  a $A = \{\text{integers greater or equal to } -4\}$ **b** $D = \{\text{integers between } -2 \text{ and } 5 \text{ inclusive}\}$

  c $C = \{-2, -1, 0, 1, 2\}$ **d** $D = \{\text{even integers between 2 and 10}\}$

1 One girl can wash two cars in 90 minutes.

  a How many girls will it take to wash four similar cars in 45 minutes?

  b How many similar cars can ten girls wash in three hours?

2 Two men in a food factory can check 120 tubs of humous for quality control in 5 minutes. Copy and complete the following table.

Number of men, m	Number of humous tubs, h	Time, t mins
2	120	5
4		10
	1080	15
m	h	
	h	t
m		t

3 Convert the following into fractions.

  a $0.\dot{5}$ **b** $0.0\dot{3}\dot{7}$

  c $0.5\dot{8}2\dot{1}$ **d** $1.05\dot{4}5\dot{6}$

4 Express $0.\dot{7}8\dot{9} \div 0.\dot{9}8\dot{7}$ as a fraction.

5 The variable y is directly proportional to x, and $y = 24$ when $x = 3$. Find

  a the formula for y in terms of x

  b the value of y when $x = 10$

  c the value of x when $y = 40$.

6 x varies directly as the cube root of y. If $x = 12$ when $y = 27$, find

  a the formula for y in terms of x **b** the value of y when $x = 32$.

7 y is inversely proportional to x, and $y = 16$ when $x = 3$. Find

  a the formula for y in terms of x

  b the value of y when $x = 8$

  c the value of x when $y = 12$.

8 A machine produces coins, of a fixed thickness, from a given volume of metal. The number of coins, N, produced is inversely proportional to the square of the diameter, d. If 4000 coins of diameter 1.5 cm are made, find

  a the formula for N in terms of d

  b the number of coins that can be produced of diameter 2 cm

  c the diameter if 1000 coins are produced.

9 The surface area, A m², of an inflatable toy is proportional to the square of its height, h m. The surface area $= 60$ m² when the height $= 2$ m. Find

  a the formula for A in terms of h

  b the value of A when $h = 3$ m

  c the value of h when $A = 540$ m².

10 The wind speed, w (miles per hour) in Florida during the hurricane season is monitored t hours after midnight on 1st January, such that w in terms of t is given by the equation $w = t^3 - 6t^2 + 4t + 25$, valid for $0 \leqslant t \leqslant 6$.

  a Copy and complete the table below.

t	0	1	2	3	4	5	6
w	25						

  b Draw the graph of w against t and use it to find

    i the wind speed at 04:30.   **ii** the time at which the windspeed is 20 mph.

11 The population of a colony of woodlice, w, t months after 1st January is given by the equation $p = \dfrac{2000}{t}$, valid for $0 \leqslant t \leqslant 6$.

  a Copy and complete the table below

t	1	2	3	4	5	6
w	2000					

  b Draw the graph of w against t and use it to find

    i the population on 15th April

    ii the date when the population is reduced by 60% from what it was on 1st February.

12 Solve the following to find the angles x, y and z.

  a

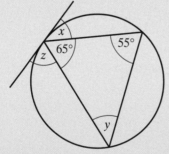

  b

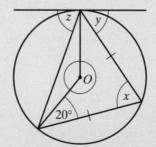

13 Solve the following to find x.

  a

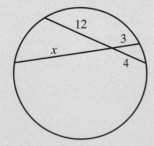

  b

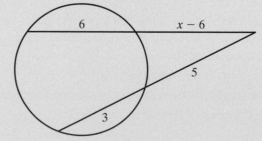

14 In a group of 100 pupils, 40 play bowls, 55 play golf, 30 do karate and 8 do nothing. 12 pupils play bowls and golf, 18 play golf and do karate, whilst 10 do karate and bowls. Use a Venn diagram to find out how many pupils

    a only do karate   **b** do only bowls and golf   **c** do all three sports.

15 On copies of the following Venn Diagram, shade the following sets.

    a $(A \cap B)'$

    b $(A' \cup B)'$

    c $(A \cap B')'$

    d $(A \cup B)'$

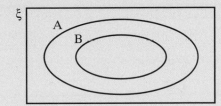

Number 2

Negative and fractional indices

Exercise 6

Simplify these

1 $a^m \times a^n$

2 $a^m \div a^n$

3 $(a^m)^n$

4 a^0

Without a calculator express the following as fractions in their simplest terms.

5 10^{-2}

6 10^{-3}

7 2^{-3}

8 3^{-2}

9 5^{-2}

10 5^{-3}

11 $(3^{-1})^3$

12 $(2^{-3})^2$

13 $3^{-2} \times 3^{-2}$

14 $4^{-1} \times 4^{-3}$

15 $5^{-2} \times 5^{-1}$

16 $6^{-5} \times 6^3$

17 $5^2 \div 5^5$

18 $10^{-1} \div 10^3$

19 $2^{-3} \div 2^2$

20 $7^2 \div 7^4$

Simplify these.

21 $a^3 \times a^{-1} \times a^4$

22 $b^2 \times b^3 \times b^{-5}$

23 $c^{-5} \times c^7 \times c^3$

24 $d^7 \times d^5 \times d^{-10}$

25 $e^7 \times e^3 \div e^6$

26 $f^5 \times f^4 \div f^7$

27 $g^{-2} \times g^{-5} \div g^{-10}$

28 $h^3 \div h^{-5} \times h^7$

For Questions 29–52, find the value of x.

29 $\sqrt{a} = a^x$

30 $\dfrac{1}{a} = a^x$

31 $\dfrac{1}{a^2} = a^x$

32 $\dfrac{1}{a^3} = a^x$

33 $\sqrt[3]{a} = a^x$

34 $\dfrac{1}{\sqrt{a}} = a^x$

35 $\dfrac{1}{\sqrt[3]{a}} = a^x$

36 $\dfrac{1}{\sqrt[4]{a}} = a^x$

37 $\sqrt{9} = 3^x$

38 $9^{\frac{1}{x}} = 3$

39 $8^{\frac{1}{x}} = 2$

40 $125^{\frac{1}{x}} = 5$

41 $2^x = \dfrac{1}{32}$

42 $4^x = \dfrac{1}{256}$

43 $5^x = \dfrac{1}{625}$

44 $7^x = \dfrac{1}{343}$

45 $125^{\frac{1}{3}} = 5^x$

46 $216^{\frac{1}{3}} = (6^2)^x$

47 $729^{\frac{1}{3}} = (3^x)^2$

48 $10\,000^{\frac{1}{4}} = (10^{2x})^2$

49 $\left(\dfrac{2}{3}\right)^x = \sqrt{\dfrac{16}{81}}$

50 $\left(\dfrac{3}{5}\right)^x = \sqrt{\dfrac{81}{625}}$

51 $\left(\dfrac{2}{3}\right)^{2x} = \sqrt{\dfrac{81}{16}}$

52 $(\sqrt{x})^{\frac{1}{2}} = 1000^{\frac{1}{3}}$

Exercise 6 ★

Evaluate the following, showing each step of your working.

1 $9^{-\frac{1}{2}}$ **2** $16^{-\frac{1}{2}}$ **3** $25^{-\frac{1}{2}}$

4 $36^{-\frac{1}{2}}$ **5** $8^{-\frac{1}{3}}$ **6** $27^{-\frac{1}{3}}$

7 $64^{-\frac{1}{3}}$ **8** $125^{-\frac{1}{3}}$ **9** $8^{-\frac{2}{3}}$

10 $27^{-\frac{2}{3}}$ **11** $64^{-\frac{2}{3}}$ **12** $125^{-\frac{2}{3}}$

13 $16^{-\frac{3}{4}}$ **14** $32^{-\frac{3}{5}}$ **15** $10\,000^{-\frac{3}{4}}$

16 $25^{-\frac{3}{2}}$ **17** $\left(\frac{4}{25}\right)^{-\frac{3}{2}}$ **18** $\left(\frac{16}{169}\right)^{-\frac{3}{2}}$

19 $\left(\frac{8}{125}\right)^{-\frac{2}{3}}$ **20** $\left(\frac{81}{625}\right)^{-\frac{3}{4}}$ **21** $\left(1\frac{7}{9}\right)^{-\frac{1}{2}}$

22 $\left(1\frac{9}{16}\right)^{-\frac{1}{2}}$ **23** $\left(11\frac{1}{9}\right)^{-\frac{1}{2}}$ **24** $\left(3\frac{3}{8}\right)^{-\frac{2}{3}}$

Simplify these.

25 $a^{\frac{1}{2}} \times a^{-\frac{5}{2}}$ **26** $b^{\frac{1}{3}} \times b^{-\frac{4}{3}}$ **27** $c^{\frac{2}{3}} \div c^{-\frac{4}{3}}$

28 $d^{-\frac{3}{4}} \div d^{-\frac{11}{4}}$ **29** $\left(e^{\frac{2}{3}}\right)^{-\frac{3}{2}}$ **30** $\left(f^{\frac{3}{5}}\right)^{-\frac{10}{3}}$

31 $\left(g^{-\frac{1}{2}}\right)^{\frac{2}{3}}$ **32** $\left(h^{-\frac{4}{5}}\right)^{-\frac{5}{2}}$

Solve the following to find the values of x and y.

33 $(a^{x}b^{y})^{\frac{2}{3}} = a^{\frac{1}{5}} \times b^{-\frac{3}{5}}$ **34** $\sqrt[4]{a^{6}b^{-4}} = a^{2x} \times b^{2y-1}$

35 $\sqrt[3]{\dfrac{a^{x}}{b^{y}}} = a^{\frac{2}{3}} \times b^{\frac{1}{5}}$ **36** $\sqrt[4]{\dfrac{a^{x+1}}{b^{y+1}}} = a^{\frac{2}{5}} \div b^{\frac{5}{6}}$

Algebra 2

Quadratic equations and quadratic inequalities

Exercise 7

Solve these equations by factorising:

1 $x^2 - 4x + 3 = 0$ **2** $x^2 + 7x + 12 = 0$ **3** $x^2 - 2x + 1 = 0$

4 $x^2 - 3x - 10 = 0$ **5** $x^2 + x - 6 = 0$ **6** $x^2 - 2x - 8 = 0$

7 $x^2 + 10x + 21 = 0$ **8** $x^2 - 7x - 18 = 0$ **9** $x^2 - 5x - 24 = 0$

10 $x^2 + 2x + 1 = 0$ **11** $x^2 + 4x = 0$ **12** $x^2 - 7x = 0$

Solve these equations by factorising.

(Hint: take out a number factor first.)

13 $2x^2 + 6x + 4 = 0$ **14** $3x^2 + 3x - 18 = 0$ **15** $2x^2 + 14x + 20 = 0$

16 $4x^2 - 8x - 60 = 0$ **17** $3x^2 - 18x + 27 = 0$ **18** $5x^2 + 20x - 60 = 0$

19 $7x^2 + 7x = 0$ **20** $2x^2 - 8x = 0$ **21** $5x^2 - 20 = 0$

22 $3x^2 - 27 = 0$

Solve these equations by using the quadratic formula, giving your answers to 3 significant figures.

23 $x^2 + 4x + 1 = 0$ **24** $x^2 - 5x + 3 = 0$ **25** $x^2 - 2x - 5 = 0$

26 $x^2 + 3x - 3 = 0$ **27** $4 + 6x + x^2 = 0$ **28** $2 - 7x + x^2 = 0$

29 $2x^2 - 6x - 7 = 0$ **30** $5x^2 + 12x + 5 = 0$ **31** $4x^2 + 10x - 3 = 0$

32 $3x^2 - 21x - 5 = 0$ **33** $x^2 + 3.7x + 1.3 = 0$ **34** $4.1x^2 - 1.2x - 0.9 = 0$

For Questions 35–40, form a quadratic equation and then solve.

35 The length of a rectangle is 1 m greater than its width and the area is $7\,m^2$. Find the dimensions of the rectangle.

36 The base of a triangle is 3 cm more than the height, and the area is $25\,cm^2$. Find the height of the triangle.

37 A right angled triangle has sides of length x, $x + 1$ and $x + 2$. Find the value of x.

38 The sum of the squares of two consecutive numbers is 145. Find the numbers.

39 A piece of wire 60 cm long is bent to form the perimeter of a rectangle of area $210\,cm^2$. Find the dimensions of the rectangle.

40 The chords of a circle intersect as shown. Use the intersecting chords theorem to find x.

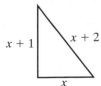

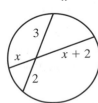

For Questions 41–50, solve the inequalities by first sketching the graph of the quadratic function.

41 $x^2 \leqslant 4$ **42** $x^2 > 49$ **43** $2x^2 < 32$

44 $3x^2 \geqslant 27$ **45** $x^2 + x - 2 < 0$ **46** $x^2 + 6x + 8 > 0$

47 $x^2 - 5x - 6 \leqslant 0$ **48** $x^2 - 7x + 12 > 0$ **49** $x^2 - 4x < 5$

50 $x^2 + 2 \geqslant 3x$

Exercise 7★

Solve the following quadratic equations by factorising.

1 $x^2 + 7x + 10 = 0$ **2** $x^2 + 2x - 15 = 0$ **3** $x^2 - 4x + 4 = 0$

4 $x^2 - 8x + 12 = 0$ **5** $x^2 - 10x + 9 = 0$ **6** $x^2 + 5x - 14 = 0$

7 $x^2 - 2x - 48 = 0$ **8** $x^2 + 15x + 36 = 0$ **9** $x^2 - 8x + 16 = 0$

10 $x^2 - 19x = 0$ **11** $4 + 3x - x^2 = 0$ **12** $18 - 3x - x^2 = 0$

Solve the following quadratic equations by factorising.

(Hint: take out a number factor first.)

13 $2x^2 + 10x + 12 = 0$ **14** $3x^2 + 3x - 6 = 0$ **15** $5x^2 + 35x + 60 = 0$

16 $4x^2 - 8x \quad 32 - 0$ **17** $7x^2 - 28x + 28 = 0$ **18** $3x^2 + 15x - 72 = 0$

19 $9x^2 + 18x = 0$ **20** $6x^2 - 18x = 0$ **21** $3x^2 - 108 = 0$

22 $7x^2 - 343 = 0$

Solve these equations by using the quadratic formula, giving your answers to 3 significant figures.

23 $x^2 - 6x + 7 = 0$ **24** $2x^2 + 8x - 3 = 0$

25 $2x^2 - 6x - 3 = 0$ **26** $3x^2 - 7x - 1 = 0$

27 $3x^2 + 5x = 10$ **28** $3x^2 = 7x + 2$

29 $4 + 13x + x^2 = 0$ **30** $10 + 3x - 2x^2 = 0$

31 $x(x + 1) + (x - 1)(x + 2) = 1$ **32** $x^2 + 62.2x + 248.6 = 0$

33 $1.2x^2 + 3.5x - 1.5 = 0$ **34** $2.5x^2 + 6.5x + 0.9 = 0$

For Questions 35–40, form a quadratic equation and then solve.

35 The sum of the squares of two consecutive odd numbers is 202. Find the numbers.

36 When a sheet of A4 paper is cut in half as shown, the result is a sheet of A5 paper. The two rectangles formed by A4 and A5 paper are similar. Find x.

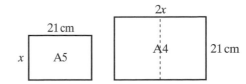

37 The hypotenuse of a right-angled triangle is 4 cm long, and the perimeter is 9 cm. Find the lengths of the other two sides.

38 The chords of a circle intersect as shown. Use the intersecting chords theorem to find x.

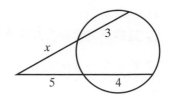

39 A circular fish pond of radius 2 m is surrounded by a path of constant width. The area of the path is the same as the area of the pond. Find the width of the path.

40 Aran spent $1200 on holiday. If he had spent $20 less each day he would have been able to stay an extra three days. How long was his holiday?

For Questions 41–50, solve the inequalities by first sketching the graph of the quadratic function.

41 $4x^2 \geqslant 64$

42 $7x^2 < 28$

43 $x^2 - 8x + 12 \leqslant 0$

44 $x^2 + 2x - 15 > 0$

45 $x^2 + 9x + 8 < 0$

46 $x^2 - 5x - 14 \leqslant 0$

47 $x^2 + 2x \geqslant 63$

48 $x^2 - 3x < 40$

49 $x^2 + 2x - 5 > 0$

50 $x^2 - 2x - 6 \leqslant 0$

Graphs 2

Using graphs to solve equations

Exercise 8

1 Draw an accurate graph of $y = x^2$ for $-3 \leqslant x \leqslant 3$. Use this graph to solve the following equations. Give the equation of the lines you use to do this.

   a $x^2 - 7 = 0$ **b** $x^2 - x = 0$ **c** $x^2 - x - 3 = 0$

   d $x^2 + x - 4 = 0$ **e** $x^2 - 2x + 1 = 0$ **f** $x^2 + 2x - 1 = 0$

2 Draw an accurate graph of $y = x^2 + x - 1$ for $-4 \leqslant x \leqslant 3$. Use this graph to solve the following equations. Give the equation of the lines you use to do this.

   a $x^2 + x - 1 = 0$ **b** $x^2 + x = 0$ **c** $x^2 + x - 2 = 0$

   d $x^2 + x - 4 = 0$ **e** $x^2 + 3x - 1 = 0$ **f** $x^2 - 1 = 0$

3 If the graph of $y = x^2 - x$ has been drawn, find the equation of the lines that should be drawn on this graph to solve the following equations.

   a $x^2 - x - 2 = 0$ **b** $x^2 - 2x = 0$

   c $x^2 - 1 = 0$ **d** $x^2 - 3x + 3 = 0$

4 If the graph of $y = x^2 + 2x - 3$ has been drawn, find the equation of the lines that should be drawn on this graph to solve the following equations.

   a $x^2 + 2x - 3 = 0$ **b** $x^2 + 2x = 0$

   c $x^2 + 2x - 6 = 0$ **d** $x^2 + x - 3 = 0$

5 Find the equation in x solved by the intersection of the following pairs of graphs.

   a $y = x^2 - x + 4$ and $y = 3 - x$ **b** $y = 2x^2 + 3x - 4$ and $y = 2x - 1$

   c $y = x^2 - 5x + 4$ and $y = 1 - x$ **d** $y = -2x^2 + x + 5$ and $y = 3x + 1$

6 Draw an accurate graph of $y = x^3$ for $-2 \leqslant x \leqslant 2$. Use this graph to solve the following equations. Give the equation of the lines you use to do this.

   a $x^3 - 2x = 0$ **b** $x^3 + 3x = 0$

   c $x^3 + x - 4 = 0$ **d** $x^3 - 3x + 1 = 0$

7 Draw an accurate graph of $y = \dfrac{3}{x}$ for $-3 \leqslant x \leqslant 3$. Use this graph to solve the following equations. Give the equation of the lines you use to do this.

   a $\dfrac{3}{x} + 2 = 0$ **b** $\dfrac{3}{x} - x = 0$

   c $\dfrac{3}{x} + x - 4 = 0$ **d** $\dfrac{3}{x} - x - 1 = 0$

8 Use a graphical method to solve the simultaneous equations $y = x^2 + 1$ and $y = x + 2$.

9 Use a graphical method to solve the simultaneous equations $y = 5 - x^2$ and $y = 2 - x$

10 A slate is blown off a roof by a high wind. The equation of the roof is $y = 4 - x$ and the path of the slate is given by $y = 4 + \dfrac{x}{2} - x^2$. Find, using a graphical method, the coordinates of where the slate lands on the roof.

Exercise 8★

1 Draw an accurate graph of $y = x^2 - 3x + 1$ for $-2 \leqslant x \leqslant 5$. Use this graph to solve the following equations. Give the equation of the lines you use to do this.

   a $x^2 - 3x + 1 = 0$ **b** $x^2 - 3x - 3 = 0$ **c** $x^2 - 4x + 1 = 0$

   d $x^2 - 4x - 1 = 0$ **e** $x^2 - 2x - 5 = 0$ **f** $x^2 - x - 3 = 0$

   g The equation $x^2 - 3x = k$ has only one solution. Find the value of k.

2 If the graph of $y = 2x^2 + 3x - 1$ has been drawn, find the equation of the lines that should be drawn on this graph to solve the following equations.

   a $2x^2 + 3x = 0$ **b** $2x^2 + 3x - 7 = 0$ **c** $2x^2 + 2x - 1 = 0$

   d $2x^2 + x - 3 = 0$ **e** $2x^2 + 7x - 4 = 0$ **f** $4x^2 + 6x - 1 = 0$

3 If the graph of $y = x^2 - 3x + 2$ has been drawn, find the equation of the lines that should be drawn on this graph to solve the following equations.

   a $x^2 - 3x = 0$ **b** $x^2 - 3x + 1 = 0$

   c $x^2 - 4x + 2 = 0$ **d** $x^2 - 2x - 2 = 0$

4 Find the equation in x solved by the intersection of the following pairs of graphs.

   a $y = 5x^2 + 13$ and $y = 3x - 4$ **b** $y = 4x^2 + 3x - 5$ and $y = 7x + 2$

   c $y = 2x^2 - 7x + 5$ and $y = x^2 + 2$ **d** $y = 2x - 1 - 3x^2$ and $y = 4 - x$

5 Draw an accurate graph of $y = 2x - x^3$ for $-2 \leqslant x \leqslant 2$. Use this graph to solve the following equations. Give the equation of the lines you use to do this.

   a $2x - x^3 = 1$ **b** $3x - x^3 = 0$

   c $x - x^3 = 1$ **d** $x^3 + x^2 = 2x + 1$

   e The equation $2x - x^3 = k$ has only one solution. Find the range of possible values for k.

6 Draw an accurate graph of $y = x + \dfrac{1}{x}$ for $-4 \leqslant x \leqslant 4$. Use this graph to solve the following equations. Give the equation of the lines you use to do this.

   a $x + \dfrac{1}{x} = 3$ **b** $\dfrac{1}{x} - x = 0$

   c $\dfrac{x}{2} + \dfrac{1}{x} + 2 = 0$ **d** $x^2 + x + \dfrac{1}{x} = 4$

7 Use a graphical method to solve the simultaneous equations

   $y = \dfrac{1}{x} - x$ and $y = x^2 - x - 3$ (Use $-3 \leqslant x \leqslant 3$)

8 Use a graphical method to solve the simultaneous equations

   $y = x^3 - 2x$ and $y = 0.5x^2 + x - 1$ (Use $-2 \leqslant x \leqslant 2$)

9 Tamas has just caught a heavy fish. The shape of his fishing rod is given by $y = 3.5x - x^2$ and the fishing line is given by $x + 2y = 8$. Use a graphical method to find the coordinates of the point where the line joins the rod. (Use $0 \leqslant x \leqslant 10$)

10 A food parcel has just been dropped by a low flying airplane flying over sloping ground. The path of the food parcel is given by $y = 40 - 0.005x^2$ and the slope of the ground is given by $y = 0.2x$. Use a graphical method to find the coordinates of the point where the food parcel will land. (Use $0 \leqslant x \leqslant 100$)

Shape and space 2

Converting measurements, circle sectors and arcs, areas and volumes of solids, areas and volumes of similar solids

Exercise 9

1 Change
  a 5 km to mm
  b 8 km to cm
  c 2×10^6 cm to km
  d 4×10^8 mm to km

2 Change
  a 5×10^6 cm^2 to m^2
  b 3 km^2 to m^2
  c 4 m^2 to mm^2
  d 2×10^{12} mm^2 to km^2

3 Change
  a 2 m^3 to cm^3
  b 4 km^3 to m^3
  c 8×10^{10} mm^3 to m^3
  d 7×10^7 cm^3 to m^3

4 Calculate the area and perimeter of the following shapes.

a

b

c

d

5 Calculate the area and perimeter of the following shapes.

a
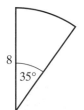

b
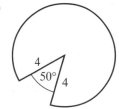

c
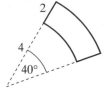

d
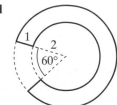

6 Find the surface area and volume of the following objects.

a

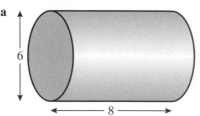

6

8

b

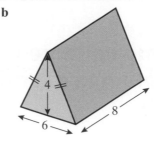

4

6

8

7 The circumference of a circle is 12 cm. Find the radius and area.

8 The volume of a spherical exercise ball is 400 cm³. Find the radius and surface area.
(For a sphere $V = \frac{4}{3}\pi r^3$, $A = 4\pi r^2$)

9 The two triangles are similar. Find the perimeter of the smaller triangle and the area of the larger triangle.

4
Area = 12

5
Perimeter = 25

10 The two shapes are similar. Find the length marked x.

8

x

Area = 100

Area = 225

11 The two cylinders are similar. Find the volume of the larger cylinder and the surface area of the smaller cylinder.

5

8

Volume = 60   Surface area = 160

12 The two yoghurt pots are similar. Find the height of the larger pot and the surface area of the smaller pot.

5

Volume = 100   Volume = 800
Surface area = 480

Exercise 9 ★

1 Change these.
  a 8.3 km to mm
  c 1.9×10⁴ m to km

  b 5.7×10^8 cm to km
  d 3.5×10^{10} mm to km

2 Change these.
  a 500 m² to cm²
  c 3.8 km² to mm²

  b 9.6×10^{15} mm² to km²
  d 8.4×10^5 mm² to m²

3 Change

    a 53 km³ to mm³

    c 10^8 mm³ to m³

    b 3.3×10^{18} mm³ to km³

    d 1000 cm³ to m³

4 Calculate the area and perimeter of the following shapes.

    a

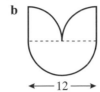

    b

    c

    d

5   **a** If the perimeter = 14 cm, find x and the area.

      b If the area = 35 cm², find x and the perimeter.

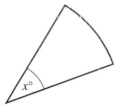

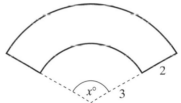

6 Find the surface area and volume of the following objects. All dimensions are in centimetres.

    a   hole $r = 3$ depth = 2

                      $r = 6$

    b

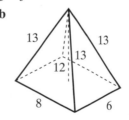

7 The perimeter of a semicircle is 24 cm. Find the radius and area.

8 The surface area of a solid hemisphere is 76 cm². Find the radius and volume.

    (For a sphere $V = \frac{4}{3}\pi r^3$, $A = 4\pi r^2$)

9 The two triangles are similar. Find the perimeter of the smaller triangle and the area of the larger triangle.

5 cm   7 cm

Area = 16 cm²   Perimeter = 28 cm

10 The two shapes are similar. Find the length marked x.

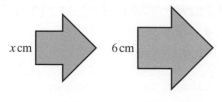

Area = 90 cm²   Area = 160 cm²

11 The two packets of cereal are similar. Find the volume of the smaller packet and the surface area of the larger packet.

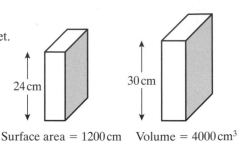

Surface area = 1200 cm   Volume = 4000 cm³

12 The volume of the Earth is 1.08×10^{12} km³ and its diameter is 12 740 km. The volume of the Moon is 2.2×10^{10} km³ and its surface area is 3.8×10^{7} km². Assuming the Earth and Moon are similar, calculate the diameter of the Moon and the surface area of the Earth.

Handling data 2

Compound probability

Exercise 10

1 A fair spinner is spun twice.
   Use a tree diagram to calculate the probability of obtaining

   a two prime numbers

   b a prime and a non-prime number.

   c an odd and an even number in that order.

   (Spinner: 4 25%, 1 25%, 3 25%, 2 25%)

2 The probability of Maria serving an ace in a game of tennis is given as $\frac{1}{15}$.
   She serves twice in succession.

   a Draw a tree diagram showing all the possibilities.

   b Find the probability of Maria serving

      i two aces   ii no aces   iii one ace.

3 Box A contains apples, $\frac{1}{4}$ of which are rotten. Box B contains lemons, $\frac{4}{5}$ of which are good. A box is randomly selected and a fruit is randomly taken.

   a Draw a tree diagram showing all the possibilities.

   b Find the probability of obtaining a good apple.

   c Find the probability of obtaining a bad lemon.

4 The probability of a baby Giant Panda being female is $\frac{4}{7}$. A family of Giant Pandas contains two offspring.

   a Draw a tree diagram showing all the possibilities.

   b Find the probability of both babies being male.

   c Find the probability of the family containing one male and one female baby.

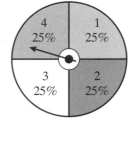

5 Two bags contain a very large number of microchips. Bag I has 10% which are faulty, whilst Bag II has 80% good microchips. Quality control randomly select a bag and then randomly remove two microchips.

   a Draw a tree diagram illustrating all possible outcomes.

   b Find the probability that the two microchips removed are

      i both faulty from Bag I

      ii both good from Bag II

      iii good and bad from Bag II.

6 Ivor has a $\frac{1}{7}$ probability of catching a cold. If he catches a cold the probability that he catches a cough is $\frac{1}{3}$. If he does not catch a cold the probability of him not catching a cough is $\frac{3}{4}$.

   a Draw a tree diagram showing all possible outcomes for Ivor.

   b Find the probability that Ivor

      i has a cold and a cough   ii only has a cough   iii has one of the two.

Exercise 10★

1 Stella is a skilled and consistent archer. The probability of her hitting the bullseye on the target for her first two shots can be found from the table below.

	1st shot	2nd shot
Hits	$\frac{2}{3}$	
Misses		$\frac{1}{5}$

   a Copy and complete the table and use it to draw a tree diagram to show all possible outcomes.

   b Find the probability of Stella scoring

      i two bullseyes   **ii** one bullseye   **iii** at least one bullseye.

2 Mrs Smith has an equal chance of producing a boy or a girl.

   a Draw a tree diagram to show all possible outcomes for a family of three children.

   b Find the probability of Mrs Smith's three children containing

      i two girls   **ii** one girl   **iii** at least one boy.

3 Scarlett considers herself a lucky hockey captain if she wins the pre-match toss on at least two occasions out of three. Is she justified in this opinion of her personal luck?

4 A toy box contains three teddy bears and two kangaroos. Jasper randomly selects one and this is *not* replaced before another is randomly taken out.

   a Draw a tree diagram to show all possible outcomes for the two selections.

   b Find the probability that Jasper's two toys are

      i two teddy bears   **ii** a teddy bear and a kangaroo   **iii** at least one kangaroo.

5 Bag A contains a large number of Christmas lights which are red, green and gold in the ratio $1:2:3$ respectively. Bag B also contains a large number of Christmas lights which are red, green and gold in the ratio $1:1:2$ respectively. Helga takes a light randomly from Bag A and she then takes one randomly from Bag B.

   a Draw a tree diagram to show all possible outcomes for the two selections.

   b Find the probability that Helga's two lights are

      i both gold   **ii** green and gold   **iii** at least one red.

6 Box X contains two white and three black marbles. Box Y contains three white and two black marbles.

Box X   Box Y

A marble is randomly taken from Box X and placed into Box Y before two marbles are randomly taken from Box Y without replacement.

   a Draw a tree diagram showing all possible outcomes for the two selections from Box Y.

   b Find the probability that the two marbles from Box Y consist of

      i two whites   **ii** a black and a white   **iii** at least one black.

Examination Practice 2

1 Simplify these, giving your answers in index form.

  a $x^2 \times x^{-1}$ **b** $x^2 \div x^{-1}$

  c $x^{\frac{1}{2}} \times x^{-\frac{1}{2}}$ **d** $x^{\frac{1}{2}} \div x^{-\frac{1}{2}}$

2 Simplify these, giving your answers in index form.

  a $(x^{-1})^2$ **b** $(x^2)^{-1}$

  c $(x^{\frac{1}{2}})^2$ **d** $(x^{-\frac{1}{2}})^{-2}$

3 Simplify these, showing sufficient working to justify your answer.

  a $8^{\frac{2}{3}}$ **b** $4^{\frac{3}{2}}$

  c $27^{-\frac{2}{3}}$ **d** $16^{-\frac{3}{2}}$

4 $x = 3 \times 10^{2n}$, $y = 6 \times 10^n$. Find simplified expressions **in standard form** for

  a xy **b** $x \div y$

5 Solve these quadratic equations.

  a $x^2 - 16 = 0$ **b** $x^2 - 16x = 0$ **c** $x^2 + 16x = 0$

6 Solve these quadratic equations by factorising.

  a $x^2 + 7x + 12 = 0$ **b** $x^2 - 4x + 3 = 0$

  c $x^2 - 2x - 8 = 0$ **d** $x^2 + 4x - 5 = 0$

7 Solve these quadratic equations by factorising.

  a $x^2 - 6x + 9 = 0$ **b** $2x^2 - 8 = 0$ **c** $2x^2 - 4x = 0$

8 Solve these quadratic equations by factorising.

  a $3x^2 - 3x - 6 = 0$ **b** $2x^2 + 5x + 2 = 0$ **c** $3x^2 - 10x + 3 = 0$

9 Solve these quadratic equations, giving your answers to 3 significant figures.

  a $x^2 + 6x + 1 = 0$ **b** $x^2 - 5x + 2 = 0$

  c $x^2 - 2x - 2 = 0$ **d** $3x^2 - 8x + 2 = 0$

10 The chords of a circle intersect as shown. Find the value of x.

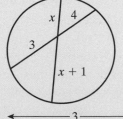

11 A flag maker is designing a large flag for a football match.
The flag is 3 m by 2 m, and the arms of the cross are of width x m.

  a Show that the area of the cross is $5x - x^2$

New regulations demand that the area of the cross is half the
area of the flag.

  b Show that this means x must satisfy the equation $x^2 - 5x + 3 = 0$

  c Solve this equation to find x correct to 3 significant figures.

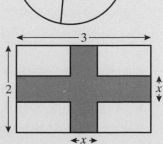

12 Solve the following quadratic inequalities.

a $x^2 \leqslant 100$

b $3x^2 > 48$

c $x^2 - x - 2 < 0$

d $x^2 + 2x - 15 \geqslant 0$

13 The graph of $y = x^2 + 1$ has been drawn. What lines should be drawn on this graph to solve the following equations?

a $x^2 = 5$

b $x^2 - 2x = 0$

c $x^2 + x - 1 = 0$

14 The diagram shows the graphs of $y = x - \frac{1}{x}$ and $y = 1 - 2x$.

What quadratic equation in x is solved by the x-coordinates of the points of intersection of these two graphs?

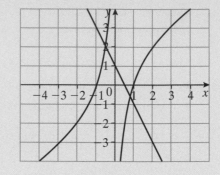

15 Change these.

a 30 km to mm

b 10^7 m^2 to km^2

c 12 m^3 to cm^3

16 A shape is made up of part of a circle of radius 3 cm and a square of side 3 cm. The circle's centre is at the corner of the square. Find the perimeter and area of the shape.

17 The diagram shows a small cylindrical tank 20 cm long part filled with a liquid. The liquid surface AB is 8 cm wide and is 3 cm below the centre.

a Calculate the radius of the cylinder.

b Calculate the angle marked x.

c Calculate the area of the sector AOB.

d Hence calculate the volume of liquid in the tank.

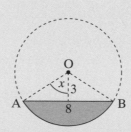

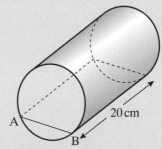

18 Nina is making three different sizes of candle for Valentine's Day. The candles are heart shaped prisms and are all similar.

The smallest candle is 5 cm tall and has a volume of 40 cm^3. The medium sized candle is 7.5 cm tall.

a Find the volume of the medium sized candle.

Nina wants the largest candle to have a volume of 400 cm^3 so it will burn for ten times longer than the smallest candle.

b Find the height of the largest candle.

The area of the heart shaped face of the medium sized candle is 27 cm^2.

c Find the areas of the heart shaped faces of the other two candles.

19 A biased spinner is marked with the scores 1, 2, 3, 4 and 5. The table shows the probabilities of these scores.

Score	1	2	3	4	5
Probability	0.05	x	0.15	$2x$	0.2

  a The spinner is spun once. Work out the probability that

    i the score is 2 or 4   **ii** the score is 2.

  b The spinner is now spun twice. Work out the probability that

    i the scores add up to 10   **ii** the scores add up to 4.

20 Darren is taking his driving test. The probability that he will pass the theory test is $\frac{3}{5}$. If he passes the theory test, he then takes the practical test. The probability that he will pass the practical test is $\frac{1}{3}$. He must pass both tests to pass the driving test.

  a Draw a probability tree diagram to represent this information.

  b Hence calculate the probability that Darren fails his driving test.

  If Darren passes his driving test, he can take a further advanced test to reduce the cost of his insurance. The probability that he passes this advanced test is $\frac{1}{4}$.

  c Calculate the probability that Darren gets reduced insurance.

Number 3

Financial arithmetic

Exercise 11

For Questions 1–6 use the following currency conversion table.

Country	Currency	Rate/UK £1
Australia	Dollars	2.51
Brazil	Reais	4.19
China	Yuan	15.28
India	Rupees	86.72
Malaysia	Ringitts	6.87
Russia	Rubles	52.12
USA	Dollars	1.96

1 Convert £500 into Australian dollars.

2 Convert 500 Australian dollars into British pounds.

3 Convert 1000 Rupees into Yuan.

4 Convert 100 Reais into Rubles.

5 Bud has 10 000 American dollars and travels through Malaysia, India and China. He wishes to convert all his money to the country's currency through which he is travelling in the ratio of 1 : 1 : 3 respectively. How much of each currency will he have?

6 How many British pounds is a Rupee millionaire worth?

7 Find the simple interest on £1000 at 5% p.a. for
   a 5 years   b 10 years   c 20 years.

8 Find the compound interest on £1000 at 5% p.a. for
   a 5 years   b 10 years   c 20 years.

9 Lena buys a lap-top computer for £550 before V.A.T. is added at 17.5%. How much does she pay?

10 Zac pays £750 for a new set of golf clubs after V.A.T. is added at 17.5%. How much do the clubs cost before V.A.T. is added?

11 Hector employs a bricklayer who charges £20/hr, a plumber who charges £30/hr and a labourer who charges £15/hr. He uses each of them for 40 hours. How much will they charge him given that V.A.T. will be added to his bill at 17.5%?

12 Gerard earns a salary of €45 000 p.a. Marcelle earns a basic rate of €20/hr and works a 40 hour week with 5% of this time added at an overtime rate of 1.5 × basic rate. Who earns more per week and by how much?

For Questions 13 and 14 use the following table which shows the tax on earned income in the Republic of Lexica.

Tax Rate	Earned Income
0%	€0–€5000 p.a.
20%	€5001–€70 000 p.a.
45%	Over €70 000 p.a.

13 Sienna is a hairdresser who earns €25 000 p.a. in Lexica. Find

   a her net salary p.a. after tax has been deducted

   b her net pay per week

   c the percentage of her gross salary which is paid in income tax.

14 Dexter is a tailor who earns €35/hr in Lexica and works a 35 hour week. Find

   a his net salary p.a. after tax has been deducted

   b his net pay per week

   c the percentage of his gross salary which is paid in income tax.

Exercise 11 ★

1 Sean buys a motorcycle for $15 000 and it depreciates in value at 9% p.a.
   How much is it worth after

   a 1 year   b 5 years   c 10 years?

2 Kristina purchases a gold brooch for $4500 which appreciates in value at 3% p.a.
   How much is it worth after:

   a 1 year   b 5 years   c 10 years?

3 Saskia has her car serviced. She is told that her bill not including V.A.T at 17.5% is £580. She is then charged £691.50. Saskia says she has been overcharged. Is Saskia correct?

4 Roman is a plumber who charges £45/hr + V.A.T. at 17.5% for his services. He charges a client £1269. How many hours did Roman work?

5 Frida is a decorator who charges £30/hr basic rate + V.A.T. at 17.5% for her services. She works a 20 hour week and charges 1.5 times her basic rate for overtime.
   She charges Mr Calderwood £1233.75 for one week's work. How many hours overtime did Frida work?

6 Rowen buys the following items for his photographic business inclusive of VAT at 17.5%.
       Digital Camera: £3500
       Tripod: £140
       Flourescent Lights: £350
       Printer: £575
   Calculate how much Rowen pays in V.A.T.

7 Kariba buys a new sports car which costs her €40 000. She pays a 20% deposit and pays the remainder over 48 months. The charge on the loan is calculated as simple interest at 9% p.a. Calculate

  a Kariba's monthly repayment

  b the total cost of the car.

8 Klaus builds an extension to his house which costs him €60 000. He pays a 25% deposit and pays the remainder over 60 months. The charge on the loan is calculated as simple interest at 7% p.a. Calculate

  a Klaus's monthly repayment

  b how much Klaus would have saved had he paid the whole amount without a loan.

For Questions 9 and 10 use the following table which shows the tax on earned income in the Republic of Lexica.

Tax Rate	Earned Income
0%	€0–€5000 p.a.
20%	€5001–€70 000 p.a.
45%	Over €70 000 p.a.

9 Davina is a super-model who pays €1million income tax in Lexica in a tax year. Calculate her gross earning before tax p.a.

10 Marco is a professional footballer in Lexica and earns €25 000 per week. Find

  a his net salary p.a. after tax has been deducted

  b the percentage of his gross salary which is paid in income tax.

Algebra 3

Simultaneous equations and functions

Exercise 12

1 Solve the following simultaneous equations.

  a $y = x^2$ and $y = x + 2$ **b** $y = x^2$ and $y = 3 - 2x$

  c $y = x^2 - 3$ and $y = 5 - 2x$

2 Solve the following simultaneous equations, giving your answers correct to 3 significant figures.

  a $y = x - 1$ and $y = x^2 + 2x - 4$ **b** $x = 2y + 1$ and $x = y^2 - 3y + 5$

3 Solve the following simultaneous equations.

  a $x^2 + y = 8$ and $x + y = 2$ **b** $x^2 + xy = 3$ and $x - y = 5$

4 Solve the following simultaneous equations, giving your answers correct to 3 significant figures.

  a $x^2 + y^2 = 7$ and $x + y = 3$ **b** $x^2 + 3y^2 = 2$ and $x + y = 1$

5 A slate is blown off a roof by a high wind. The equation of the line of the roof is $y = 4 - x$ and the path of the slate is given by $y = 4 + \frac{x}{2} - x^2$. Find the coordinates of where the slate leaves and lands on the roof.

6 If $f(x) = 7 - 2x$, calculate   **a** $f(3)$ **b** $f(-2)$ **c** $f(0)$

7 If $g : x \rightarrow 5x - 4$, calculate x if   **a** $g(x) = 6$ **b** $g(x) = -4$ **c** $g(x) = -1$

8 If $f(x) = 2x + 3$, find   **a** $f(x) - 1$ **b** $f(x - 1)$

9 If $f : x \rightarrow 3(x - 1)$ and $g : x \rightarrow 4x + 2$, find the value of x such that $f(x) = g(x)$.

10 If $f(x) = x^2$ and $g(x) = 6x - 8$, find the values of x such that $f(x) = g(x)$.

11 State which values of x cannot be included in the domain of:

  a $f(x) = \dfrac{1}{x + 2}$ **b** $g(x) = \dfrac{5}{1 - 2x}$

  c $h(x) = \sqrt{x - 1}$ **d** $h(x) = \sqrt{x + 1}$

12 Find the range of the following functions:

  a $f(x) = x - 1$ **b** $g(x) = x^2 - 1$ **c** $h(x) = (x - 1)^2$

13 $f(x) = 2x + 1$ and $g(x) = \sqrt{x}$

  a Find   **i** $fg(x)$ **ii** $gf(x)$

  b What values should be excluded from the domain of **i** $fg(x)$ **ii** $gf(x)$?

  c Find and simplify $ff(x)$.

14 Find the inverse of

   a $f(x) = 5x - 4$ **b** $g(x) = 3 - x$

   c $h(x) = \dfrac{1}{x + 3}$ **d** $f(x) = x^2 - 3$

15 $f(x) = 4 - 5x$ and $g(x) = \dfrac{4 - x}{5}$

   a Find the functions   **i** $fg(x)$ **ii** $gf(x)$

   b Describe the relationship between functions f and g.

   c Write down the exact value of $fg(\pi)$.

Exercise 12★

1 Solve these simultaneous equations.

   a $y = 8 - x$ and $y = x^2 - 4$ **b** $2x = 3y$ and $x^2 + y^2 = 13$ **c** $x + y = 5$ and $xy = 6$

2 Solve these simultaneous equations.

   a $x + 2y = 3$ and $x^2 + 2y^2 = 3$ **b** $x + y = 2$ and $3x^2 - y^2 = 2$

3 Solve the following simultaneous equations, giving your answers correct to 3 significant figures.

   a $2x + y = 4$ and $x^2 + xy = 2$ **b** $x - y = 2$ and $x^2 + xy - y^2 = 5$

4 Solve the following simultaneous equations, giving your answers correct to 3 significant figures.

   a $x + y = 5$ and $\dfrac{6}{y} - \dfrac{6}{x} = 4$ **b** $x - y = 2$ and $\dfrac{x}{y} + \dfrac{y}{x} = 3$

5 A food parcel has just been dropped by a low flying airplane flying over sloping ground. The path of the food parcel is given by $y = 40 - 0.005x^2$ and the slope of the ground is given by $y = 0.2x$. Find the coordinates of where the food parcel will land.

6 Tamas has just caught a heavy fish. The shape of his fishing rod is given by $y = 3.5x - x^2$ and the fishing line is given by $x + 2y = 8$. Find the coordinates of where the line joins the rod.

7 If $f(x) = x - \dfrac{1}{x}$, calculate   **a i** $f(2)$ **ii** $f(-2)$

   b Why can't $f(0)$ be calculated?

8 If $g : x \rightarrow 5x - x^2$, calculate x if   **a** $g(x) = 6$ **b** $g(x) = -6$

9 If $f(x) = 4 + 3x$, find   **a** $2f(x)$ **b** $f(2x)$

10 If $f(x) = x^2 + 5x + 3$ and $g(x) = (2x + 1)^2$, find the values of x such that $f(x) = g(x)$.

11 State which values of x cannot be included in the domain of

   a $f(x) = \dfrac{1}{3 + 4x}$ **b** $g(x) = \dfrac{7}{(x - 1)^2}$ **c** $h(x) = \sqrt{5 - 2x}$

12 Find the range of the following functions

   a $f(x) = 3x^2 - 1$ **b** $g(x) = (x - 4)^2$ **c** $h(x) = \sqrt{5 - 2x}$

13 $f(x) = \sqrt{x + 1}$ and $g(x) = 3x + 2$

  a Find **i** $fg(x)$ **ii** $gf(x)$

  b What values should be excluded from the domain of **i** $fg(x)$ **ii** $gf(x)$?

  c Find and simplify $gg(x)$.

14 a Find the inverse of

   i $f(x) = 7(2 - x)$ **ii** $g(x) = \sqrt{x + 5}$ **iii** $h(x) = (x + 1)^2$ **iv** $f(x) = \frac{4}{x} - 2$

  b $f(x) = \dfrac{x}{x - 1}$

   i Find the inverse $f^{-1}(x)$.

   ii Hence describe the relationship between $f(x)$ and $f^{-1}(x)$.

15 $f(x) = \dfrac{1}{(x + 1)}$ and $g(x) = \dfrac{1}{x} - 1$

  a Find the functions **i** $fg(x)$ **ii** $gf(x)$

  b Hence describe the relationship between functions f and g.

  c Write down the exact value of $fg(\sqrt{19})$.

Graphs 3

Tangents to a curve

Exercise 13

For Questions 1–4 refer to the graph shown.

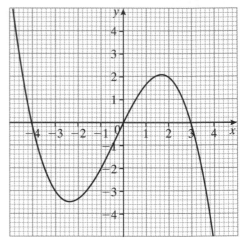

1   Find the gradient when

    a $x = 1$ **b** $x = 3$ **c** $x = -2$

2   Where is the gradient equal to

    a 1   **b** −1   **c** 0?

3   The tangents at the point P and Q are parallel to
the line $y = x$. Find the coordinates of P and Q.

4   The tangents at the points R and S are parallel to
the line $x + y = 2$. Find the coordinates of R and S.

5   Plot an accurate graph of $y = \frac{1}{2}x^2$ for $-4 \leqslant x \leqslant 4$. By drawing suitable tangents, find the gradient of
the graph at

    a i $x = 1$ **ii** $x = -2$ **iii** $x = 3$

    b Where on the curve is the gradient equal to -2?

    c Find the equation of the tangent to the curve at the point where $x = 2$.

6   The distance-time graph shows a bee's journey from its hive.

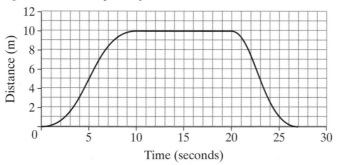

    a Find the velocity of the bee at

        i 5 s   **ii** 15 s   **iii** 22 s

    b When was the bee flying fastest?

    c Describe the bee's journey.

7   The velocity v m/s of a skydiver for the first 5 seconds of her fall is given by $v = 10t - t^2$.

    a Draw an accurate graph of v against t for $0 \leqslant t \leqslant 5$.

    b Find the acceleration of the skydiver after **i** 1 s   **ii** 3 s

    c When was the skydiver's acceleration 5 m/s²?

8   Mala fills a flower vase with water to a depth of 20 cm. The level of water drops every day by 15% due to evaporation.

   a   Copy and complete the table, and use it to plot the graph of depth against time for $0 \leqslant t \leqslant 8$.

t (days)	0	1	2	3	4	5	6	7	8
d (cm)	20		14.45						5.45

   b   How many days does it take for the water level to drop by half?

   c   Find the rate in cm per day that the water is dropping when $t = 2$.

   d   Find the rate in **cm per hour** that the water is dropping when $t = 6$.

   e   What is the quickest rate in cm per day at which the water drops, and when does this occur?

Exercise 13★

For Questions 1–4 refer to the graph shown.

1   a   Find the gradient when
      i $x = 1$ ii $x = -1$ iii $x = 4$
   b   Where is the gradient equal to
      i -1 ii   2   iii   0?

2   The curve passes through the point $(3, -1.4)$.
   a   Find the gradient at this point.
   b   Find the equation of the tangent at this point.

3   The line $y = mx + 3$ is a tangent to the curve. Find the two possible values of m.

4   The line $y = k - x$ is a tangent to the curve. Find the two possible values of k.

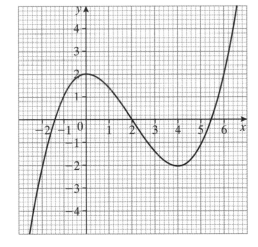

5   a   Plot an accurate graph of $y = 4 - \frac{1}{2}x^2$ for $-4 \leqslant x \leqslant 4$. By drawing suitable tangents, find the gradient of the graph at
      i $x = -1$ ii $x = 2$ iii $x = -3$
   b   What do you think the gradient at $x = 10$ will be?
   c   What is the connection between the gradient at $x = a$ and $x = -a$?
   d   Find the equation of the tangent to the curve at the point where $x = -2$.
   e   The line $y = mx + 5$ is a tangent to the curve. Find the values of m.

6   Galileo is rolling a ball down a slope. He finds that the distance, d m, after t seconds is given by $d = 2.5t^2$.
   a   Plot an accurate graph of d against t for $0 \leqslant t \leqslant 4$.
   b   Find the velocity, v m/s, after   i   1 s   ii   2 s   iii   3 s   iv   4 s   v   0 s
   c   Plot a graph of v against t.
   d   What does your graph tell you about the acceleration?

7 Pedro sets off on a run. His distance, d m, from his starting point after t seconds is given by
$d = \dfrac{t^3}{5} - 2t^2 + 5t$ for $0 \leqslant t \leqslant 8$.

  a Plot an accurate graph of distance against time for $0 \leqslant t \leqslant 8$.

  b Describe Pedro's run for $0 \leqslant t \leqslant 8$.

  c What was his velocity after i 1 second ii 3 seconds?

  d When was his velocity 6 m/s?

8 A population of eagles is increasing by 50% every 10 years.

  a Copy and complete the table and use it to plot the graph of N against t.

t (years)	0	10	20	30	40	50	60	70
N (eagles)	100		225					

  b Estimate the rate of growth of the population in eagles **per year** after

  i 30 years   ii 50 years

  c When was the rate of growth 20 eagles per year?

Shape and space 3

Vectors

Exercise 14

1 Given that $\mathbf{u} = \begin{pmatrix} 1 \\ 2 \end{pmatrix}$, $\mathbf{v} = \begin{pmatrix} -2 \\ 3 \end{pmatrix}$ express **p**, **q**, **r** and **s** as column vectors and find their magnitudes where

  a $\mathbf{p} = \mathbf{u} + \mathbf{v}$

  b $\mathbf{q} = \mathbf{u} - \mathbf{v}$

  c $\mathbf{r} = 2\mathbf{u} + 3\mathbf{v}$

  d $\mathbf{s} = 3\mathbf{u} - 2\mathbf{v}$

2 Given that $\mathbf{u} = \begin{pmatrix} 3 \\ 1 \end{pmatrix}$, $\mathbf{v} = \begin{pmatrix} 2 \\ 5 \end{pmatrix}$ and $\mathbf{w} = \begin{pmatrix} 0 \\ -4 \end{pmatrix}$ express **p**, **q**, **r** and **s** as column vectors and find their magnitudes where

  a $\mathbf{p} = \mathbf{u} + \mathbf{v} + \mathbf{w}$

  b $\mathbf{q} = \mathbf{u} - \mathbf{v} - \mathbf{w}$

  c $\mathbf{r} = 2\mathbf{u} + 3\mathbf{v} + 4\mathbf{w}$

  d $\mathbf{s} = 2(\mathbf{u} - 2\mathbf{v} + 3\mathbf{w})$

3 Esther walks a route described using column vectors **p** and **q**, where the units are in km. $\mathbf{p} = \begin{pmatrix} 1 \\ 2 \end{pmatrix}$ and $\mathbf{q} = \begin{pmatrix} -3 \\ 5 \end{pmatrix}$. Her journey from O is given by vector **w** where $\mathbf{w} = 2\mathbf{p} + 3\mathbf{q}$. Find

  a the vector **w**

  b the magnitude and bearing of **w**

  c the speed of Esther's journey from O if she takes 4 hours to complete it.

4 OAB is a triangle such that $\overrightarrow{OA} = \mathbf{a}$ and $\overrightarrow{OB} = \mathbf{b}$ and M is the mid-point of AB. Express the following in terms of **a** and **b**

  a $\overrightarrow{AB}$

  b $\overrightarrow{AM}$

  c $\overrightarrow{OM}$

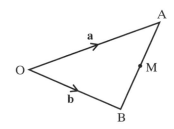

5 OAB is a triangle such that $\overrightarrow{OA} = 2\mathbf{x}$ and $\overrightarrow{OB} = 2\mathbf{y}$ and $AM : MB = 1 : 1$. Express the following in terms of **x** and **y**

  a $\overrightarrow{AB}$

  b $\overrightarrow{AM}$

  c $\overrightarrow{OM}$

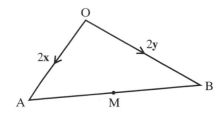

6 ABCDEF is a regular hexagon with centre O. $\overrightarrow{AB} = \mathbf{p}$ and $\overrightarrow{BC} = \mathbf{q}$.
Express the following in terms of $\mathbf{p}$ and $\mathbf{q}$

a $\overrightarrow{ED}$ **b** $\overrightarrow{DE}$

c $\overrightarrow{AC}$ **d** $\overrightarrow{AE}$

Exercise 14★

1 $\mathbf{p} = \begin{pmatrix} 3 \\ 2 \end{pmatrix}$ and $\mathbf{q} = \begin{pmatrix} 3 \\ -5 \end{pmatrix}$. Find

a i $\mathbf{p} + \mathbf{q}$ **ii** $2\mathbf{p} - \mathbf{q}$

b the values of constants m and n such that $m\mathbf{p} + n\mathbf{q} = \begin{pmatrix} 12 \\ -13 \end{pmatrix}$

c the values of constants r and s such that $\mathbf{p} + r\mathbf{q} = \begin{pmatrix} s \\ -8 \end{pmatrix}$

d the values of constants u and v such that $u(\mathbf{p} + \mathbf{q}) + v(2\mathbf{p} - \mathbf{q}) = \begin{pmatrix} 0 \\ 21 \end{pmatrix}$.

2 Given that $\mathbf{a} = \begin{pmatrix} 1 \\ 2 \end{pmatrix}$, $\mathbf{b} = \begin{pmatrix} 4 \\ 3 \end{pmatrix}$ and $m\mathbf{a} + n\mathbf{b} = \begin{pmatrix} -2 \\ 1 \end{pmatrix}$ where m and n are constants find the values of m and n.

3 The position vector $\mathbf{r}$ of a comet with respect to an origin O, t seconds after being detected is given by $\mathbf{r} = \begin{pmatrix} 1 \\ 5 \end{pmatrix} + t\begin{pmatrix} 2 \\ -1 \end{pmatrix}$, where the units are in km.

a Copy and complete the table for position vector $\mathbf{r}$.

t	0	1	2	3	4	5
$\mathbf{r}$	$\begin{pmatrix} 1 \\ 5 \end{pmatrix}$	$\begin{pmatrix} 3 \\ 4 \end{pmatrix}$				

b Plot the path of the comet for $0 \le t \le 5$.

c Calculate the speed of the comet in km/h and its bearing.

4 OPQ is a triangle. OR : RP = 2 : 1 and S is the mid-point of OQ. M is the mid-point of PQ. If $\overrightarrow{OR} = \mathbf{r}$ and $\overrightarrow{OS} = \mathbf{s}$, express the following in terms of $\mathbf{r}$ and $\mathbf{s}$

a $\overrightarrow{RS}$ **b** $\overrightarrow{OP}$

c $\overrightarrow{PQ}$ **d** $\overrightarrow{OM}$

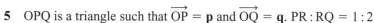

5 OPQ is a triangle such that $\overrightarrow{OP} = \mathbf{p}$ and $\overrightarrow{OQ} = \mathbf{q}$. PR : RQ = 1 : 2

a Express in terms of vectors $\mathbf{p}$ and $\mathbf{q}$

  i $\overrightarrow{PQ}$ **ii** $\overrightarrow{PR}$ **iii** $\overrightarrow{OR}$.

b If $\overrightarrow{OS} = k\overrightarrow{OR}$, where k is a constant and OS : SR = 3 : 2, find

  i k **ii** $\overrightarrow{OS}$ in terms of $\mathbf{p}$ and $\mathbf{q}$.

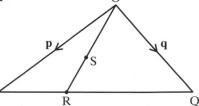

6 OPQ is a triangle such that $\overrightarrow{OP} = \mathbf{p}$ and $\overrightarrow{OQ} = \mathbf{q}$. OM : MP = 3 : 2 and $PN = \frac{2}{5}PQ$.

a Express in terms of vectors $\mathbf{p}$ and $\mathbf{q}$

  i $\overrightarrow{MP}$ **ii** $\overrightarrow{PQ}$ **iii** $\overrightarrow{PN}$ **iv** $\overrightarrow{MN}$

b How are OQ and MN related?

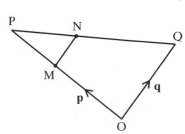

Handling data 3

Histograms

Exercise 15

1 In an experiment the lengths of 80 daisy stalks were measured. The results are shown in the table.

  a Construct a histogram for these results.

  b Calculate an estimate of the number of daisies in this group that have a stalk length between 4 and 8 mm.

  c Calculate an estimate of the mean length of the daisy stalks.

Length, l (mm)	Frequency, f
$0 < l \leqslant 5$	5
$5 < l \leqslant 10$	10
$10 < l \leqslant 20$	22
$20 < l \leqslant 30$	25
$30 < l \leqslant 50$	18

2 40 people were asked how long it takes them to travel to work. The results are shown in the table.

  a Construct a histogram for these results

  b Calculate an estimate of the number of people who took more than 60 minutes to travel to work.

  c Calculate an estimate of the number of people who took less than 24 minutes to travel to work.

Time, t (mins)	Frequency, f
$15 < t \leqslant 20$	12
$20 < t \leqslant 30$	12
$30 < t \leqslant 40$	10
$40 < t \leqslant 70$	6

3 A farmer checks the masses of a sample of apples for quality control. The unfinished table and histogram show the results.

Mass, m (g)	Frequency, f
$60 < m \leqslant 70$	
$70 < m \leqslant 80$	22
$80 < m \leqslant 100$	40
$100 < m \leqslant 120$	20
$120 < m \leqslant 160$	

  a Use the histogram to complete the table.

  b Use the table to complete the histogram.

  c Calculate an estimate of the number of apples with mass between 75 and 95 grams.

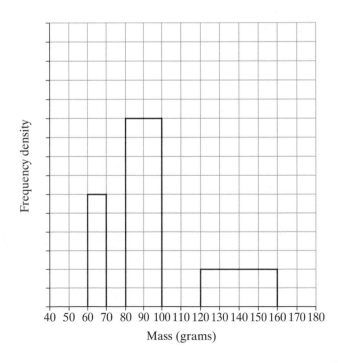

45

4 The unfinished table and histogram give the waiting times
   of 100 patients at a doctor's surgery. The number of patients
   waiting 10–12 minutes is 8 more than the number waiting
   0–6 minutes.

Time t mins	Frequency
$0 < t \leqslant 6$	
$6 < t \leqslant 10$	26
$10 < t \leqslant 12$	
$12 < t \leqslant 14$	18
$14 < t \leqslant 22$	24

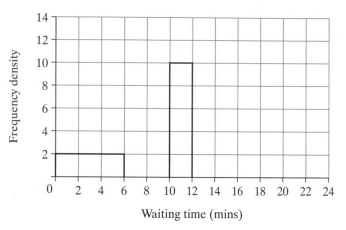

a Use the histogram to complete the table.

b Use the table to complete the histogram.

c Calculate an estimate of the percentage of patients who wait between 8 and 16 minutes.

d Calculate an estimate of the median waiting time.

Exercise 15★

1 In an experiment the lengths of 100 cats' whiskers were
   measured. The results are shown in the table.

   a Construct a histogram for these results.

   b Calculate an estimate of the number of whiskers in this
      group that have a length between 5.5 and 8.5 mm.

   c Calculate an estimate of the mean length of the whiskers.

Length, l (mm)	Frequency, f
$5 < l \leqslant 6$	18
$6 < l \leqslant 6.5$	15
$6.5 < l \leqslant 7$	21
$7 < l \leqslant 8$	26
$8 < l \leqslant 10$	20

2 The table gives the times of 50 cyclists in a race.

   a Construct a histogram of these results.

   b Calculate an estimate of the percentage of cyclists
      in this group that had a time of less than 48 minutes.

   c Calculate an estimate for the median time.

Time, t (mins)	Frequency, f
$20 < t \leqslant 30$	5
$30 < t \leqslant 35$	4
$35 < t \leqslant 40$	9
$40 < t \leqslant 50$	22
$50 < t \leqslant 70$	7
$70 < t \leqslant 100$	3

3 A retailer checks the lifetimes of a sample
of 200 Christmas tree lights for quality
control. The unfinished table and histogram
show the results.

Life, t (hrs)	Frequency, f
$60 < t \leqslant 80$	
$80 < t \leqslant 90$	26
$90 < t \leqslant 95$	20
$95 < t \leqslant 100$	25
$100 < t \leqslant 115$	
$115 < t \leqslant x$	51

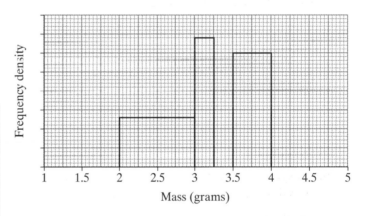

a Use the histogram to complete the table.

b Calculate the value of x.

c Use the table to complete the histogram.

4 The unfinished table and histogram
give the birth masses of 200 babies.
The difference between the number
of babies in the 2–3 kg class and
the 3–3.25 kg class is 18.

Mass m kg	Frequency
$2 < m \leqslant 3$	
$3 < m \leqslant 3.25$	
$3.25 < m \leqslant 3.5$	30
$3.5 < m \leqslant 4$	
$4 < m \leqslant 4.75$	

a Use the histogram to complete the table.

b Use the table to complete the histogram.

c Calculate the probability that the birth mass is between 2.5 and 4.5 kg.

d Calculate an estimate of the median mass.

Examination Practice 3

1 A television costs $380 plus 17.5% VAT. Calculate the total price.

2 A firm buys a computer for £1116.25 including 17.5% VAT. The firm can reclaim the VAT. How much can it reclaim?

3 Keighly is paid €12 per hour for a 35 hour week, with overtime paid at €18 per hour. One week she works for 42 hours. How much does she get paid?

4 Solve the following simultaneous equations.

   a $x - 2y = 2$ and $x + y^2 = 5$ **b** $2x^2 - y^2 = 8$ and $x + y = 2$

5 $f(x) = 3 - 5x$.

   a Find   **i** $f(2)$ **ii** $f(-2)$

   b Find the inverse function $f^{-1}(x)$.

   c Solve the equation $f(x) = f^{-1}(x)$.

6 **a** Draw the graph of $y = x^2$ for $-3 \leqslant x \leqslant 3$

   b Find the range of   **i** $f(x) = x^2$ **ii** $f(x) = x^2 + 1$ **iii** $f(x) = x^2 - 1$

7 $f(x) = 3 - x$

   a Find $ff(x)$ in its simplest form.

   b What does your answer to part **a** show about the function f?

8 $\mathbf{p} = \begin{pmatrix} 3 \\ 4 \end{pmatrix}$, $\mathbf{q} = \begin{pmatrix} -2 \\ 1 \end{pmatrix}$

   a Express the following as column vectors as simply as possible

     i $\mathbf{p} + \mathbf{q}$ **ii** $3\mathbf{p} - 2\mathbf{q}$

   b Calculate the magnitude of $\mathbf{p} + \mathbf{q}$.

   c Calculate the direction of $3\mathbf{p} - 2\mathbf{q}$ as a bearing.

9 Copy and complete the following table showing the time in minutes that a group of 90 school children spent reading on a particular day.

Time, t (mins)	$10 < t \leqslant 20$	$20 < t \leqslant 25$	$25 < t \leqslant 30$	$30 < t \leqslant 40$	$40 < t \leqslant 50$	$50 < t \leqslant 70$	$70 < t \leqslant 95$
Frequency	7	15	25		12		
Frequency density			5	1.8		0.4	

10 The table shows the tax rates for the land of Oz.

   a Calculate the tax paid on a salary of $25 000 per annum.

   b Mrs Read pays $5 000 per annum tax. What is her salary?

   c Mr Singh pays $15 000 per annum tax. What is his salary?

Tax rate		Band of taxable income per annum
Personal allowance	0%	$0–$10 000
Basic rate	20%	$10 001–$60 000
Higher rate	40%	Over $60 000

   d Mrs Tams is paid $90 000 per annum. She receives a 10% pay rise. By what percentage does her pay increase by after tax has been deducted?

11 $f(x) = 1 - \sqrt{x}, g(x) = 2 + 3x$

  a Which values of x cannot be included in the domain of $f(x)$?

  b Find

    i $fg(x)$ **ii** $gf(x)$

  c Solve

    i $fg(x) = g(x)$ **ii** $gf(x) = g(x)$

12 $f(x) = (2 - x)^2$. Find both values of p such that $p = f(p)$

13 Ella has just done her first skydive. The graph shows how her height varied with time for the first 100 seconds after leaving the airplane.

  a How high was she after 40 s?

  b What was her velocity after

    i 5 s   **ii** 30 s   **iii** 50 s?

  c Describe the first 100 seconds of her fall.

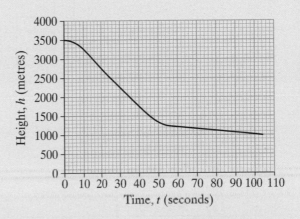

14 Zara is designing a pendant. The shape of the curved outline is given by $4x^2 + y^2 = 4$ and the line AB is given by $2x + y = 1$.
Calculate the coordinates of A and B, giving your answers to 3 significant figures.

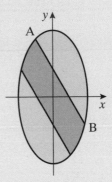

15 ABCD is a trapezium in which $\overrightarrow{AB} = \mathbf{b}$, $\overrightarrow{BC} = \mathbf{a}$ and $\overrightarrow{AD} = 2\mathbf{a}$. The points P, Q, R and S are the mid-points of the sides AB, BC, CD and AD as shown.

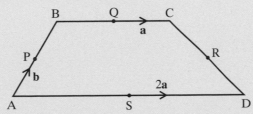

  a Find and simplify the vectors

    i $\overrightarrow{AS}$ **ii** $\overrightarrow{PQ}$ **iii** $\overrightarrow{CD}$ **iv** $\overrightarrow{SR}$

  b What can be deduced about PQ and SR?

  c What can be deduced about the quadrilateral PQRS?

16 A health farm recorded the milk consumption, m ml, of its 180 clients one day. The table and histogram show the results.

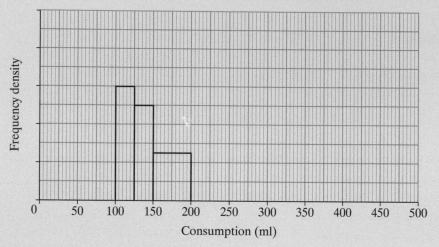

Consumption, m (ml)	Frequency, f
$0 < m \leqslant 50$	9
$50 < m \leqslant 100$	42
$100 < m \leqslant 125$	30
$125 < m \leqslant 150$	
$150 < m \leqslant 200$	
$200 < m \leqslant 300$	22
$300 < m \leqslant 500$	

a Use the histogram to complete the table.

b Use the table to complete the histogram.

c Calculate an estimate of the median amount of milk consumed that day.

d Calculate an estimate of the percentage of clients who consumed between 70 and 170 ml of milk that day.

Number 4

Irrational numbers and surds

Exercise 16

1 Which of the following are rational?

    a $0.\dot{8}$ **b** $\sqrt{17}$ **c** $\sqrt{225}$

    d 2π **e** $\frac{3}{13}$

2 Find a rational number   **a** between $\sqrt{5}$ and $\sqrt{7}$ **b** between $\sqrt{7}$ and 3.

3 Find an irrational number   **a** between 4 and 5   **b** between 4.5 and 5.

4 Square the following.

    a $\sqrt{6}$ **b** $\sqrt{a}$ **c** $3\sqrt{2}$

    d $\sqrt{2} \times \sqrt{3}$ **e** $\sqrt{2} \div \sqrt{3}$

5 Write the following as square roots of a single integer.

    a $3\sqrt{3}$ **b** $2\sqrt{7}$

    c $4\sqrt{5}$ **d** $\sqrt{2} + 2\sqrt{2}$

6 Simplify the following as far as possible.

    a $8\sqrt{5} - 4\sqrt{5}$ **b** $8\sqrt{5} + 4\sqrt{5}$

    c $8\sqrt{5} \times 4\sqrt{5}$ **d** $8\sqrt{5} \div 4\sqrt{5}$

7 Simplify the following as far as possible.

    a $\sqrt{18}$ **b** $\sqrt{20}$

    c $\sqrt{27}$ **d** $\sqrt{28}$

8 Simplify the following as far as possible.

    a $\sqrt{8} + \sqrt{18}$ **b** $\sqrt{32} - 2\sqrt{2}$ **c** $\sqrt{24} + 3\sqrt{6}$

9 Simplify the following as far as possible.

    a $\dfrac{\sqrt{75}}{\sqrt{3}}$ **b** $\dfrac{\sqrt{32}}{2}$

    c $\dfrac{\sqrt{125}}{\sqrt{5}}$ **d** $\sqrt{\dfrac{9}{25}}$

10 Simplify the following.

    a $(1 + \sqrt{7})^2$ **b** $(2 - \sqrt{7})^2$ **c** $(1 + \sqrt{7})(1 - \sqrt{7})$

11 Rationalise the denominator and simplify the following as far as possible

    a $\dfrac{8}{\sqrt{8}}$ **b** $\dfrac{12}{\sqrt{6}}$

    c $\dfrac{2\sqrt{18}}{\sqrt{12}}$ **d** $\dfrac{\sqrt{5} - 5}{\sqrt{5}}$

12 The area of a square is 27 cm^2.

    a Show that the perimeter of the square is $12\sqrt{3}$ cm.

    b Show that the length of a diagonal of the square is $3\sqrt{6}$ cm.

13 Show, by using the quadratic formula, that the solutions to $x^2 - 2x - 2 = 0$ are $x = 1 + \sqrt{3}$ or $x = 1 - \sqrt{3}$.

14 The area of a circle is 5 cm². Show that the circumference is $2\sqrt{5\pi}$ cm.

15 Show that the solution of the equation $x\sqrt{2} + x\sqrt{8} = 6$ is $x = \sqrt{2}$.

Exercise 16★

1 Which of the following are rational?

   a $0.\dot{5}\dot{3}$ **b** $\sqrt{7} \times \sqrt{7}$ **c** $\sqrt{7} + \sqrt{7}$

   d $\sqrt{729}$ **e** π^2

2 Find a rational number   **a** between $\sqrt{17}$ and $\sqrt{19}$ **b** between $\sqrt{29}$ and $\sqrt{31}$.

3 Find an irrational number   **a** between 11.5 and 11.75   **b** between π and $\sqrt{10}$.

4 Square the following.

   a $\sqrt{ab}$ **b** $\sqrt{\dfrac{a}{b}}$

   c $\dfrac{1}{2\sqrt{a}}$ **d** $\dfrac{4\sqrt{a}}{\sqrt{2b}}$

5 Rationalise the denominator and simplify the following.

   a $\dfrac{\sqrt{18}}{\sqrt{6}}$ **b** $\dfrac{4\sqrt{2}}{\sqrt{8}}$ **c** $\dfrac{10\sqrt{6}}{\sqrt{150}}$

   d $\dfrac{16}{\sqrt{8}}$ **e** $\dfrac{81}{\sqrt{27}}$

6 Simplify the following as far as possible.

   a $2\sqrt{48} + 3\sqrt{27}$ **b** $\sqrt{63} - \sqrt{28}$ **c** $\sqrt{1000} - \sqrt{40} - \sqrt{90}$

7 Simplify the following as far as possible.

   a $(\sqrt{2} + 3)(\sqrt{2} - 1)$ **b** $(3 - \sqrt{2})^2$ **c** $(\sqrt{8} + 2\sqrt{2})(2\sqrt{8} - \sqrt{2})$

8 Simplify the following as far as possible.

   a $\sqrt{(a^2b)}$ **b** $(\sqrt{a} + \sqrt{b})^2$

   c $(1 + \sqrt{a})(1 - \sqrt{a})$ **d** $(\sqrt{a} + \sqrt{a})^2$

9 Show that $\dfrac{3\sqrt{5}}{2\sqrt{7}}, \dfrac{15}{2\sqrt{35}}$ and $\dfrac{3\sqrt{35}}{14}$ all represent the same number.

10 The length of a rectangle is twice the width. The area is 196 cm². Show that

   a the perimeter is $42\sqrt{2}$ cm   **b** the length of a diagonal is $7\sqrt{10}$.

11 The area of a semi-circle is 6π cm². Show that the perimeter is $2\pi\sqrt{3} + 4\sqrt{3}$ cm.

12 Show, by using the quadratic formula, that the solutions to

   a $x^2 - 2x - 4 = 0$ are $x = 1 + \sqrt{5}$ or $x = 1 - \sqrt{5}$

   b $x^2 - 4x - 1 = 0$ are $x = 2 + \sqrt{5}$ or $x = 2 - \sqrt{5}$.

13 **a** Show that the solution of the equation $x\sqrt{12} + x\sqrt{75} = 21$ is $x = \sqrt{3}$.

   b Show that the solution of the equation $\dfrac{3x}{\sqrt{2}} + 5 = x\sqrt{8}$ is $x = 5\sqrt{2}$.

14 **a** Express $\left(\dfrac{1}{\sqrt{3}}\right)^5$ in the form $\dfrac{\sqrt{a}}{b}$ where a and b are integers.

   b Express $\left(\dfrac{4}{\sqrt{8}}\right)^3$ in the form $c\sqrt{d}$ where c and d are integers.

15 Show that $7\sqrt{1 - \frac{1}{49}} = n\sqrt{3}$, where n is an integer, whose value is to be found.

Algebra 4

Algebraic fractions and equations with fractions

Exercise 17

Simplify the following.

1 $\dfrac{9x - x^2}{x}$

2 $\dfrac{2x + 6x^2}{2x}$

3 $\dfrac{4x + 12}{x + 3}$

4 $\dfrac{x^2 + 5x}{x + 5}$

5 $\dfrac{2x^2 + 8x}{x + 4}$

6 $\dfrac{x^2 - 4}{x - 2}$

7 $\dfrac{x^2 + 5x + 4}{x + 4}$

8 $\dfrac{x^2 + x - 2}{x^2 - 6x + 5}$

Express as single fractions.

9 $\dfrac{1}{2} + \dfrac{x}{4}$

10 $\dfrac{x}{2} + \dfrac{x}{4}$

11 $\dfrac{x}{3} + \dfrac{x - 1}{6}$

12 $\dfrac{x - 1}{2} - \dfrac{x + 1}{5}$

13 $\dfrac{4x - 12}{3x} \times \dfrac{9x^2}{6x - 18}$

14 $\dfrac{x^2 - 4}{x^2 - 2x - 8} \div \dfrac{x - 2}{x - 4}$

15 $\dfrac{1}{x} + \dfrac{1}{3x}$

16 $\dfrac{3}{x} + \dfrac{4}{y}$

17 $\dfrac{1}{x - 2} - \dfrac{1}{x}$

18 $\dfrac{2}{x - 1} + \dfrac{3}{x + 2}$

19 $\dfrac{1}{x} + \dfrac{1}{x(x + 1)}$

20 $\dfrac{2x}{x - 2} - \dfrac{x}{5}$

Solve the following equations.

21 $\dfrac{x}{9} = \dfrac{2}{3}$

22 $\dfrac{x}{5} = \dfrac{x - 2}{3}$

23 $\dfrac{x}{2} = \dfrac{x - 1}{3}$

24 $\dfrac{x + 2}{4} = \dfrac{x + 1}{5}$

25 $\dfrac{x}{2} + \dfrac{x}{3} = 6$

26 $\dfrac{4x - 3}{3} + 2 = x$

27 $\dfrac{x + 4}{2} + \dfrac{3x + 1}{3} = 0$

28 $\dfrac{x + 2}{3} - \dfrac{x + 4}{5} + 6 = 0$

29 $\dfrac{4}{x} = \dfrac{1}{5}$

30 $\dfrac{-2}{3} = \dfrac{6}{x}$

31 $\dfrac{3}{x + 2} = \dfrac{5}{x}$

32 $\dfrac{4}{x - 6} = \dfrac{7}{x + 3}$

33 $\dfrac{x}{x + 2} - \dfrac{1}{x} = 1$

34 $x = 5 - \dfrac{6}{x}$

35 $\dfrac{1}{x - 2} - \dfrac{1}{x + 1} = \dfrac{1}{6}$

36 $3 + x + \dfrac{2}{x} = 0$

Exercise 17★

Simplify the following.

1 $\dfrac{(2x)^2 - 8x}{4x}$

2 $\dfrac{10x + 15x^2}{5x}$

3 $\dfrac{7x^2 - 49}{x^2 - 7}$

4 $\dfrac{6x^2 + 9x}{2x + 3}$

5 $\dfrac{5x^2 - 15x}{3 - x}$

6 $\dfrac{x^2 - y^2}{x - y}$

7 $\dfrac{x^2 + 7x - 18}{x + 9}$

8 $\dfrac{x^2 - x - 20}{x^2 - 16}$

Express as single fractions.

9 $\dfrac{x + 1}{2} - \dfrac{x - 1}{6}$

10 $\dfrac{2x - 1}{3} - \dfrac{x + 3}{4}$

11 $\dfrac{2(x + 3)}{7} - \dfrac{3(x - 3)}{5}$

12 $\dfrac{x^2 - 1}{x} \times \dfrac{x^2}{x^2 + 2x + 1}$

13 $\dfrac{x^2 - 49}{x + 3} \div \dfrac{x + 7}{x^2 - 9}$

14 $\dfrac{1}{2x} - \dfrac{1}{3x}$

15 $\dfrac{2}{x + 3} + \dfrac{3}{x - 2}$

16 $\dfrac{1}{1 - x} - \dfrac{2}{2 + x}$

17 $\dfrac{3}{2x - 1} - \dfrac{4}{3x + 1}$

18 $x + \dfrac{1}{1 + x}$

19 $\dfrac{x + 3}{x + 4} - \dfrac{x + 5}{x + 6}$

20 $\dfrac{4}{x + 3} + \dfrac{12}{x^2 - 9}$

For Questions 21–36 solve the equation.

21 $\dfrac{2x}{3} - \dfrac{x}{4} = 5$

22 $\dfrac{2x + 1}{3} = \dfrac{x}{7}$

23 $\dfrac{x - 1}{2} = \dfrac{3x + 1}{5}$

24 $10 = 1 - \dfrac{x - 3}{2}$

25 $\dfrac{x + 2}{3} = \dfrac{2x - 1}{4} - 1$

26 $\dfrac{x - 3}{11} + \dfrac{2x}{3} = 2$

27 $\dfrac{7x - 3}{4} - \dfrac{2x + 1}{5} = 0$

28 $\dfrac{x + 2}{2} + \dfrac{x + 3}{3} + \dfrac{x + 4}{4} = 0$

29 $\dfrac{3}{2x - 1} + \dfrac{5}{x} = 0$

30 $\dfrac{3}{x + 3} = \dfrac{-9}{2x - 19}$

31 $\dfrac{x - 1}{x + 2} + \dfrac{x + 3}{x - 2} = 2$

32 $\dfrac{4}{x} = \dfrac{x}{3x - 5}$

33 $\dfrac{x - 1}{x + 2} = \dfrac{3x + 2}{x}$

34 $1 + \dfrac{2}{x} = \dfrac{24}{x^2}$

35 $\dfrac{1}{x} - \dfrac{1}{x + 1} = \dfrac{1}{x + 4}$

36 $\dfrac{x}{2x - 4} - \dfrac{x}{3x - 6} = 1$

Graphs 4

Differentiation, turning points and motion of a particle

Exercise 18

1 Differentiate the following using the correct notation.

    a $y = 3x$ **b** $y = 10$ **c** $y = x^3$

    d $y = x^4$ **e** $y = x^5$ **f** $y = 2x^6$

    g $y = 3x^5$ **h** $y = 20x^8$

2 Differentiate the following using the correct notation.

    a $y = 2x^3 + 5x^2$ **b** $y = 7x^2 - 3x$ **c** $y = 1 + 5x^3$

    d $y = 3x^4 - 5x^2$ **e** $y = x^2(x + 5)$ **f** $y = (x - 3)(x + 5)$

    g $y = (2x - 1)(x - 4)$ **h** $y = (x + 2)^2$

3 Find the gradients of the tangents to the following curves at the given points.

    a $y = 2x^2$ $(1, 2)$ **b** $y = 3x - 2x^2$ $(2, -2)$

    c $y = 2x^3 + 10x^2$ $(1, 12)$ **d** $y = (x + 5)(2x + 1)$ $(2, 35)$

4 The flow, Q m³/s of a river t hours after midnight is given by the equation
    $Q = t^3 - 8t^2 + 14t + 10$.

    a Differentiate to find $\dfrac{dQ}{dt}$.

    b Find the rate of change of flow of the river at   **i** midnight   **ii** 02:00   **iii** 05:00.

5 For the curve $y = x^3 - 3x + 2$

    a find $\dfrac{dy}{dx}$

    b find the co-ordinates of the curve where $\dfrac{dy}{dx} = 0$

    c determine whether each point is a maximum or minimum.

6 For the curve $y = x^3 + 3x^2 - 9x + 1$

    a find $\dfrac{dy}{dx}$

    b find the co-ordinates of the curve where $\dfrac{dy}{dx} = 0$

    c determine whether each point is a maximum or minimum.

7 The displacement, s metres, of a ball after t seconds is given by $s = 50 + 12t^2$.

    a Find an expression for the ball's velocity, $v = \dfrac{ds}{dt}$ and find its velocity at $t = 2$.

    b Find an expression for the ball's acceleration, $a = \dfrac{dv}{dt}$ and find its acceleration at $t = 2$.

8 The displacement, s metres, of a particle after t seconds is given by $s = t^3 + 4t^2 - 3t + 1$.

    a Find an expression for the velocity, $v = \dfrac{ds}{dt}$ and find its velocity at $t = 10$.

    b Find an expression for the acceleration, $a = \dfrac{dv}{dt}$ and find its acceleration at $t = 10$.

Exercise 18★

1 Differentiate the following using correct notation.

a $y = \dfrac{1}{x}$

b $y = \dfrac{1}{x^2}$

c $y = \sqrt{x}$

d $y = x^{-3}$

e $y = 4x^{-4}$

f $y = \dfrac{1}{2x^2}$

g $y = \dfrac{2x^3 + 3x^2 + 4}{x}$

h $y = \dfrac{(x + 3)(2x - 1)}{x}$

2 Find the gradients of the tangents to the following curves at the given points.

a $y = \dfrac{2}{x}$ $(1, 2)$

b $y = 2\sqrt{x} + x$ $(1, 3)$

c $y = \left(x + \dfrac{1}{x}\right)^2$ $(1, 4)$

d $y = \dfrac{(1 - 2x)(1 + 2x)}{2x}$ $(2, -3\frac{3}{4})$

3 Find the equation of the tangent to the curve $y = x + \dfrac{1}{x}$ at the point where $x = 2$.

4 Find the equations of the tangents to the curve $y = x^2 - 3x$ at the points where it cuts the x-axis.

5 a Find the equation of the tangent to the curve $y = x^3 - 2x^2 + 1$ at the point where $x = 2$.

   b Determine the nature of any stationary points on this curve.

6 If the curve $y = 3x^3 + \dfrac{p}{x} + 3$ has a gradient of 8 where $x = 1$, find the value of p.

7 The temperature, $C°$ Centigrade, of the sea off Brighton t months after New Year's Day is given by

$C = 4t + \dfrac{16}{t}$ and is only valid for $1 \leqslant t \leqslant 6$.

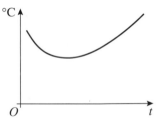

a Find $\dfrac{dC}{dt}$

b Find the coolest sea temperature in this period and when it occurs.

c Find the rate at which the sea temperature is cooling on February 1st.

8 The population, P, of a rare orchid in the Amazon rainforest over t years is given by $P = 5t^2 + \dfrac{10\,000}{t}$ and is valid for $1 \leqslant t \leqslant 5$.

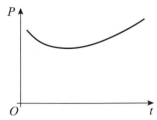

a Find $\dfrac{dP}{dt}$.

b Determine when the lowest population occurs and find its value.

9 A catapult projects a small stone vertically upwards such that its height, s m, after t seconds is given by $s = 50t - 5t^2$. Find the greatest height reached by the stone and the time it occurs.

10 The displacement of a comet, s km, after t seconds is given by $s = t(t^2 - 300)$ km.

   a Find expressions for the velocity, $v = \dfrac{ds}{dt}$ and the acceleration, $a = \dfrac{dv}{dt}$.

   b Find the velocity and acceleration of the comet after 5 seconds.

   c Find when the comet is momentarily at rest.

Shape and space 4

Trigonometry for angles up to 180°, 3D trigonometry, sine and cosine rule

Exercise 19

For Questions **1–6**, find the side marked with a letter. Write your answers correct to 3 significant figures.

1
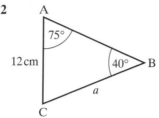

2
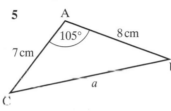

3
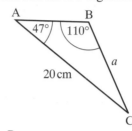

4
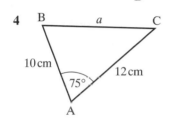

5
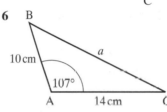

6
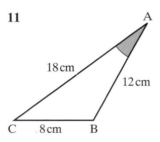

For Questions **7–12**, find angle A. Write your answers correct to 3 significant figures.

7
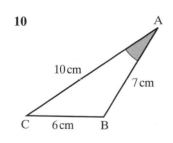

8
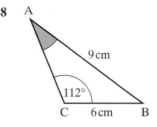

9
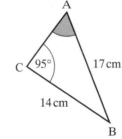

10
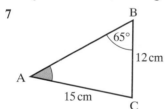

11
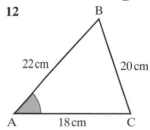

12
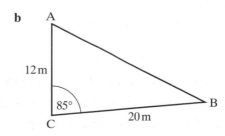

13 Find the area of each triangle ABC.

a

b

57

14 ABCDEFGH is a rectangular box. Find

a FH

b BH

c angle BHF

d the angle between plane ABGH and plane EFGH.

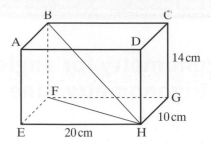

Exercise 19★

1 Find side a and angle C for

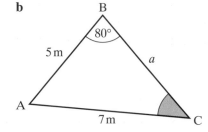

a

b

2 A yacht, Y, is 10 km from a port, P, on a bearing of 156°, whilst fishing boat F, is 20 km from the port on a bearing of 044°. Find

a the distance of the yacht from the fishing boat

b the bearing of F from Y

c the bearing of Y from F.

3 A hot-air balloon drifts 28 km from its starting point O, on a bearing of 080°, then 62 km on a bearing of 225° to reach point A. Find

a the distance OA

b the bearing of O from A.

c A truck departs from O at the same time as the balloon. It travels along OA and arrives at A at the same time as the balloon. The balloon's speed is 30 km/h.
Find the speed of the truck.

4 Find the perimeter of triangle ABC if its area is 250 m² and angle A is greater than 90°.

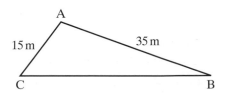

5 VABCD is a right-pyramid on a square base. Find

a the length AC

b the pyramid's height

c angle AVB

d the angle between plane AVB and base ABCD

e the total surface area of the pyramid including its base.

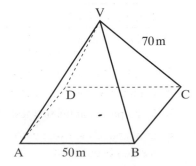

More probability

Exercise 20

1 Hamish owns a herd of cows. $\frac{4}{5}$ of the cows are Jersey whilst the rest are Friesian. He randomly picks two from the herd without replacement.

  a Calculate an estimate of the probability that he has selected
    i two Friesian cows   ii a Friesian and a Jersey cow.

  b A third cow is randomly selected from the herd. Calculate the probability that Hamish now has at least two Friesian cows from his sample of three.

2 Marco is a wine waiter who has decanted his wine in preparation for a large and important wedding. However, he has forgotten to label the bottles. He knows that the types of wine are in the numbers below.

	French	Italian
White	32	15
Red	18	45

  a If he randomly takes a bottle, calculate the probability that the wine is
    i French
    ii Italian.

  b If he randomly takes a bottle of French wine calculate the probability that the wine is
    i white
    ii red

  c In a hurry Marco randomly takes three bottles without replacement. Calculate to 3 significant figures, the probability that he has at least one bottle of Italian red wine.

3 Boris and Gunter are taking a scuba dive leader test. If they fail the first test only one re-test is permitted. Their probabilities of passing are given in the table.

	Boris	Gunter
P(pass on 1st test)	0.65	0.75
P(pass on 2nd test)	0.85	0.80

These probabilities are independent of each other. Find the probabilities that

  a Gunter becomes a dive leader at the second test

  b Boris passes at his first test whilst Gunter fails his first test

  c only one of the two men becomes a scuba dive leader, assuming that a re-test is taken if the first test has been failed

4 The probability of a red kite laying a certain number of eggs is given in the table.

Find

a the value of k

b the probability that a red kite

   i lays at least two eggs   **ii** lays at most two eggs.

c The probability that two red kites lay a total of at least three eggs.

Number of eggs	Probability
0	0.1
1	k
2	0.3
3	0.4
4	0.06
$\geqslant 5$	0.04

Exercise 20★

1 Three cards are randomly dealt to Kasper from a normal pack of 52 playing cards without replacement. Calculate the probability that his three cards contain

   a three red cards   **b** three Kings   **c** two Aces   **d** at least two Hearts.

2 Claudia, Kristina and Megan compete in the 100 m sprint and the 200 m sprint in an athletics competition. The probabilities of them *not* gaining 1st place are shown in the table.

Calculate the probability that in this athletics match

a Megan wins both races

b Kristina wins at least one of her races

c none of them win the   **i** 100 m   **ii** 200 m

d at least one of the two events is won by none of these three girls.

	100 m	200 m
Claudia	0.5	0.7
Kristina	0.7	0.8
Megan	0.9	0.7

3 A new test is 90% successful in detecting a weak heart in race horses. If a weak heart is detected, the correcting operation has an 80% success rate, the first time it is attempted. If this operation is unsuccessful, it can be repeated only twice more, with respective success rates of 60% and 40%.

a Draw a tree diagram illustrating all possible outcomes.

b Use this diagram to find the probability of a horse having a weak heart which is:

   i cured after one operation   **ii** cured at the third operation   **iii** cured.

4 A box contains 20 jewels. The jewels are either diamonds or rubies. Amrita randomly selects two jewels from the box without replacement and finds that they are both diamonds. The probability of this event happening is $\frac{21}{38}$. Calculate

a the number of diamonds in the box

b the probability that she picks out a diamond and a ruby from the original box of 20 jewels.

5 Lars has not revised for his Physics multiple choice test and decides to guess all the answers. The probability of him picking the right answer is r. The probabilty of Lars getting exactly one question correct from the first two questions is $\frac{8}{25}$. Find

a The value of r.   **b** The number of options in each question of the test.

Examination Practice 4

1 Simplify the following.

  a $5\sqrt{3} + 2\sqrt{3}$ **b** $5\sqrt{3} \times 2\sqrt{3}$ **c** $12\sqrt{3} \div 2\sqrt{3}$

  d $(3 + \sqrt{2})^2$ **e** $(3 - \sqrt{2})(3 + \sqrt{2})$ **f** $\dfrac{\sqrt{45} + \sqrt{18}}{\sqrt{5} + \sqrt{2}}$

2 Rationalise the denominator of the following.

  a $\dfrac{6}{\sqrt{2}}$ **b** $\dfrac{2}{\sqrt{18}}$ **c** $\dfrac{1}{7 - 2\sqrt{3}}$

3 Write down an irrational number between 8 and 9.

4 Simplify the following.

  a $\dfrac{3x + 24}{2x + 16}$ **b** $\dfrac{x + 5}{x^2 + 10x + 25}$

  c $\dfrac{x^2 - y^2}{2(x - y)^2}$ **d** $\dfrac{x^3 - xy^2}{x^3 + 2x^2y + xy^2}$

5 Simplify the following.

  a $\dfrac{8x + 8}{x - 5} \times \dfrac{x^2 - 5x}{4x + 4}$ **b** $\dfrac{x^2 + 10x + 21}{x^2 - 25} \times \dfrac{x - 5}{x^2 + 4x - 21}$

  c $\dfrac{x + 4}{x + 6} \div \dfrac{x^2 - x - 20}{x^2 + 4x - 12}$ **d** $\dfrac{x^2 + 2xy + y^2}{3(x^2 - y^2)} \div \dfrac{x^2 + xy}{12(x^2 - 2xy + y^2)}$

6 Express the following as a single fraction.

  a $\dfrac{5}{x} + \dfrac{7}{y}$ **b** $\dfrac{5}{x + 3} + \dfrac{3}{x - 5}$

  c $\dfrac{1}{x - 5} - \dfrac{1}{x + 2}$ **d** $\dfrac{x + 1}{x^2 + x - 30} \div \dfrac{x^2 - 1}{x - 5}$

7 Solve the following equations.

  a $\dfrac{x + 3}{x + 1} = \dfrac{x + 3}{2x - 3}$ **b** $\dfrac{3}{2} + \dfrac{3x}{x + 7} = \dfrac{12}{5}$ **c** $\dfrac{5x}{x + 3} + \dfrac{40}{3x - 1} = 10$

8 Investigate and classify the stationary points of the following curves.

  a $y = x^2 - 4x + 5$ **b** $y = x^3 - 12x + 3$ **c** $y = x^3 + 3x^2 - 9x + 1$

9 The displacement, s metres, of a train after t seconds is given by $s = 50 + 12t^2$.

  a Find an expression for the train's velocity, $v = \dfrac{ds}{dt}$ and find its velocity at $t = 3$.

  b Find an expression for the train's acceleration, $a = \dfrac{dv}{dt}$ and find its acceleration at $t = 3$.

10 The displacement of a shooting star, s km, after t seconds is given by $s = t(t^2 - 500)$ km.

  a Find expressions for the velocity, $v = \dfrac{ds}{dt}$ and the acceleration, $a = \dfrac{dv}{dt}$.

  b Find the velocity and acceleration of the shooting star after 10 seconds.

  c Find when the shooting star is momentarily at rest.

11 Calculate the length of PQ and the size of angles P and Q.

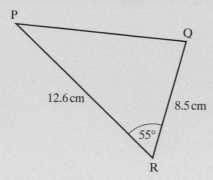

12 Ship A leaves a port at 14:00 hours and sails at a speed of 15 km/h on a bearing of 130°. Ship B leaves the same port at 14:00 hours and sails at a speed of 10 km/h on a bearing of 240°.

   a Draw a diagram to show the positions of the port and the two ships at 16:00hrs.

   b Calculate the distance between the ships at 16:00 hrs.

   c Calculate the bearing of ship A from ship B.

13 A ski-slope is shown:
Calculate

   a the angle that BF makes with the base ADEF

   b the length DF

   c the angle CF makes with the base ADEF

   d the speed of a skier in m/s travelling from M, the mid-point of BC directly to F in 30 seconds.

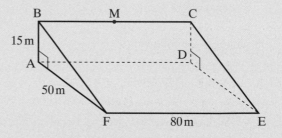

14 Igor is at the top of a 25 m tree when he sees Jesper due East of him at an angle of depression of 25°. Simultaneously he observes Karen due North at an angle of depression of 15°. How far apart are Jesper and Karen at this moment?

15 A veterinary surgeon has two independent tests, X and Y, for the presence of a virus in a turkey. Tests are performed in the order X then Y.

Probabilities of a positive test depending on the presence of the virus		
	Test X	Test Y
Virus present	$\frac{3}{5}$	$\frac{5}{6}$
Virus not present	$\frac{4}{5}$	$\frac{6}{7}$

   a Find the probability that one test will be positive and one negative on
     i an infected turkey
     ii an uninfected turkey.

   b If at least one positive test is the criterion for indicating an infected turkey, find the probability that after two tests
     i on an infected turkey, the criterion is not met
     ii on an uninfected turkey, the criterion is met.

Examination papers

Paper 1

1 The surface area of the Earth is 510 million km^2.
The surface area of the Pacific Ocean is 180 million km^2.

  a Express 180 million as a percentage of 510 million.
     Give your answer correct to 2 significant figures.

  The surface area of the Arctic Ocean is 14 million km^2.
  The surface area of the Southern Ocean is 35 million km^2.

  b Find the ratio of the surface area of the Arctic Ocean to the surface area of the Southern Ocean.
     Give your ratio in the form $1 : n$.

4400 May 2006

2 a Expand $4(3y - 2)$.

  b Expand and simplify $(x - 2)(x + 5)$.

  c Factorise $8y + 20z$.

  d Factorise $x^2 - 5x$.

  e Solve $4(3x - 2) = 28$.

3 Two points, A and B, are plotted on a centimetre grid.
A has coordinates $(2, 1)$ and B has coordinates $(8, 5)$.

  a Work out the coordinates of the mid-point of the line joining A and B

  b Use Pythagoras' Theorem to work out the length of AB.
     Give your answer correct to 3 significant figures.

4400 May 2004

4 a Solve the inequality $4x - 5 > 3$.

  b Represent the solution to part **a** on a number line.

5 Robin fires 15 arrows at a target.
The table shows information about his scores.

  a Find his median score.

  b Work out his mean score.

Score	Frequency
1	6
2	3
3	1
4	1
5	4

4400 May 2006

6 The probability that a person chosen at random has brown eyes is 0.45.
The probability that a person chosen at random has green eyes is 0.12.

a Work out the probability that a person chosen at random has either brown eyes **or** green eyes.

250 people are to be chosen at random.

b Work out an estimate for the number of people who will have green eyes.

4400 May 2005

7 The diagram shows a prism.
the cross section of the prism is a trapezium.
The lengths of the parallel sides of the trapezium are 9 cm and 5 cm.
The distance between the parallel sides of the trapezium is 6 cm.
The length of the trapezium is 15 cm.

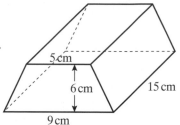

a Work out the area of the trapezium.

b Work out the volume of the prism.

4400 May 2005

8 A = {1, 2, 3, 4}
B = {1, 3, 5}

a List the members of the set   **i** A ∩ B   **ii** A ∪ B

b Explain clearly the meaning of 3 ∈ A

4400 May 2004

9 The table gives information about the ages, in years, of the 80 members of a sports club.

Age t years	Frequency
$10 < t \leqslant 20$	8
$20 < t \leqslant 30$	38
$30 < t \leqslant 40$	28
$40 < t \leqslant 50$	4
$50 < t \leqslant 60$	2

a Work out an estimate for the mean age of the 80 members.

b Complete a cumulative frequency table.

c Draw a cumulative frequency graph for your table.

d Use your graph to find an estimate for the median age of the members of the club. Show your method clearly.

4400 May 2005

10 The diagram represents part of the London Eye.
A, B and C are points on a circle, centre O.
A, B and C represent three capsules.
The capsules A and B are next to each other.
A is at the bottom of the circle and C is at the top.
The London Eye has 32 equally spaced capsules
on the circle.

a Show that the angle AOB = 11.25°.

b Find the size of the angle between BC and the
horizontal.

The capsule moves in a circle of diameter 135 m.

c Calculate the distance moved by a capsule in
making a complete revolution.
Give your answer correct to 3 significant figures.

The capsule moves at an average speed of 0.26 m/s.

d Calculate the time taken for a capsule to make a complete revolution.
Give your answer in minutes, correct to the nearest minute.

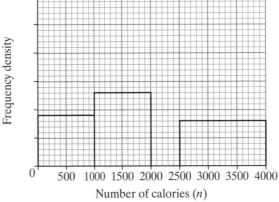

Diagram
NOT
accurately
drawn

135 m

4400 May 2006

11 a Find the gradient of the line with equation $3x - 4y = 15$

b Work out the coordinates of the point of intersection of the line with equation
$3x - 4y = 15$ and the line with equation $5x + 6y = 6$

4400 May 2004

12 $f(x) = 3x - 2$, $g(x) = \frac{1}{x}$

a Find the value of **i** $fg(3)$ **ii** $gf(3)$

b **i** Express the composite function $fg(x)$ in its simplest form as $fg(x) = \ldots$

ii Which value of x must be excluded from the domain of $fg(x)$?

c Solve the equation $f(x) = fg(x)$

13 The unfinished table and histogram show
information from a survey of women about
the number of calories in the food they eat
in one day.

Number of calories (n)	Frequency
$0 < n \leqslant 1000$	90
$1000 < n \leqslant 2000$	
$2000 < n \leqslant 2500$	140
$2500 < n \leqslant 4000$	

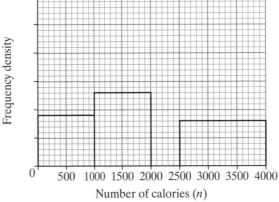

Frequency density

Number of calories (n)

a **i** Use the information in the table to
complete the histogram.

ii Use the information in the histogram to complete the table.

b Find an estimate for the upper quartile of the number of calories.
You must make your method clear.

4400 May 2004

14 An electrician has wires of the same length made from the same material.
The electrical resistance, R ohms, of a wire is inversely proportional to the square of its radius, r mm.
When $r = 2$, $R = 0.9$

   a **i** Express R in terms of r.
       ii Sketch the graph of R against r.

One of the electrician's wires has a radius of 3 mm.
   b Calculate the electrical resistance of this wire.

4400 May 2004

15 A rectangular piece of card has length $(x + 4)$
cm and width $(x + 1)$ cm.
A rectangle 5 cm by 3 cm is cut from the corner
of the piece of card.
The remaining piece of card, shown shaded in
the diagram, has an area of 35 cm².

   a Show that $x^2 + 5x - 46 = 0$

   b Solve $x^2 + 5x - 46 = 0$ to find x.
      Give your answer correct to 3 significant figures.

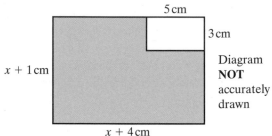

Diagram
NOT
accurately
drawn

4400 May 2005

16 Part of the graph of $y = x^2 - 2x - 4$ is shown on the grid.

   a Write down the coordinates of the minimum point of the curve.

   b Use the graph to find estimates of the solutions to the equation $x^2 - 2x - 4 = 0$.
      Give your answers correct to 1 decimal place.

   c Draw a suitable straight line on the grid to find estimates of the solutions of the equation
      $x^2 - 3x - 6 = 0$

   d For $y = x^2 - 2x - 4$
      i Find $\dfrac{dy}{dx}$
      ii find the gradient of the curve at the point where $x = 6$

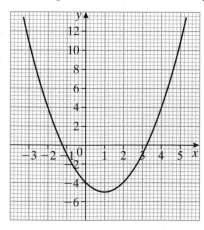

4400 May 2006

17 In the triangle shown, BC = 7.8 cm

Angle ABC = 42°

Angle BAC = 110°

   a   Calculate the length of AB.

        Give your answer correct to 3 significant figures.

   b   Calculate the area of triangle of ABC.

        Give your answer correct to 3 significant figures.

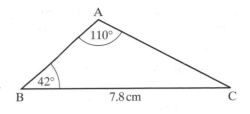

18 The diagram shows six counters.

Each counter has a letter on it.

Bishen puts the six counters in a bag. He takes a counter at random from the bag.

He records the letter which is on the counter and replaces the counter in the bag.

He than takes a second counter at random and records the letter which is on the counter.

   a Calculate the probability that the first letter will be A and the second letter will be N.

   b Calculate the probability that both letters will be the same.

4400 May 2005

19 A cylindrical tank has radius of 30 cm and a height of 45 cm.

The tank contains water to a depth of 36 cm.

A metal sphere is dropped into the water and is completely covered.

The water level rises by 5 cm.

Calculate the radius of the sphere.

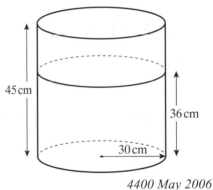

4400 May 2006

20 Simplify fully $\dfrac{3}{x+2} + \dfrac{x+17}{x^2-x-6}$

Paper 2

1 In the diagram, ABC and ADE are straight lines.
CE and BD are parallel.
AB = AD.
Angle BAD = 38°.
Work out the value of p.
Give a reason for each step in your working.

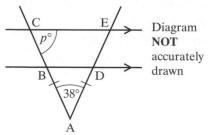

Diagram **NOT** accurately drawn

4400 May 2006

2 Krishnan used 611 units of electricity.
The first 182 units cost £0.0821 per unit.
The remaining units cost £0.0704 per unit.
Tax is added at 5% of the total amount.
Complete Krishnan's bill

182 units at £0.0821 per unit.	£.........
...... units at £0.0704 per unit	£.........
Total amount	£
Tax at 5% of the total amount	£.........
Amount to pay	£

4400 May 2005

3 Arul has x sweets.
Nikos had four times as many sweets as Arul.

a Write down an expression, in terms of x, for the number of sweets Nikos had.

Nikos gave 6 of his sweets to Arul.
Now they both have the same number of sweets.

b Use this information to form an equation in x.

c Solve your equation to find the number of sweets that Arul had at the start.

4400 May 2006

4 The mean height of a group of 4 girls is 158 cm.

a Work out the total height of the four girls.

Sarah joins the group and the mean height of the five girls is 156 cm.

b Work out Sarah's height.

4400 May 2004

5 Plumbers' solder is made from tin and lead.
The ratio of the weight of tin to the weight of lead is 1:2

a Work out the weight of tin and the weight of lead in 120 grams of plumbers' solder.

b What weight of plumbers' solder contains 25 grams of tin?

4400 May 2004

6 **a** Express 24 as the product of powers of its prime factors.

b Express 90 as the product of powers of its prime factors.

c Find the highest common factor of 24 and 90.

d Find the lowest common multiple of 24 and 90.

7 This formula is used in science. $v = \sqrt{2gh}$

   a Hanif uses the formula to work out an estimate for the value of v without using a calculator when $g = 9.812$ and $h = 0.819$

   Write down approximate values for g and h that Hanif could use.

   b Make h the subject of the formula $v = \sqrt{2gh}$

4400 May 2004

8 a The universal set, $\xi = \{\text{Angela's furniture}\}$
     $A = \{\text{Chairs}\}$
     $B = \{\text{Kitchen furniture}\}$
     Describe fully the set $A \cap B$.

   b $P = \{2, 4, 6, 8\}$
     $Q = \{\text{Odd numbers less than } 10\}$
     i List the members of the set $P \cup Q$.
     ii Is it true that $P \cap Q = \varnothing$? Explain your answer.

4400 May 2005

9 Triangle PQR is right-angled at R
  PR = 4.7 cm and PQ = 7.6 cm.

   a Calculate the size of angle PQR.
   Give your answer correct to 1 decimal place.
  The length, 7.6 cm, of PQ is correct to 2 significant figures.

   b **i** Write down the upper bound of the length of PQ.
     ii Write down the lower bound of the length of PQ.

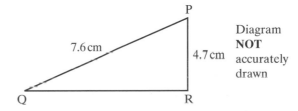

Diagram **NOT** accurately drawn

4400 May 2004

10 A mobile phone company makes a special offer.
    Usually one minute of call time costs 5 cents.
    For the special offer, this call time is increased by 20%.

   a Calculate the call time which costs 5 cents during this special offer.
   Give your answer in seconds.

   b Calculate the cost per minute for the special offer.

   c Calculate the percentage decrease in the cost per minute for the special offer.

4400 May 2006

11 a Solve $3x + 5 = 2(3x - 2)$.

   b Solve the simultaneous equations $x + 2y = -3$ and $2x - 3y = 8$.

   c Expand and simplify $(x + 3)(3x - 2)$.

12 Quadrilateral **P** is mathematically similar to quadrilateral **Q**.

   a Calculate the value of x.

   b Calculate the value of y.

   The area of quadrilateral **P** is 60 cm².

   c Calculate the area of quadrilateral **Q**.

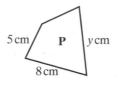 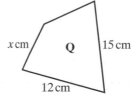

Diagram **NOT** accurately drawn

4400 May 2004

13 Two chords, AB and CD, of a circle intersect at right angles at X.

AX = 2.8 cm

BX = 1.6 cm

CX = 1.2 cm

Calculate the length AD.

Give your answer correct to 2 significant figures.

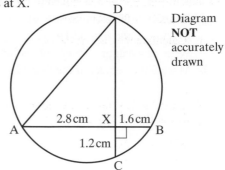

Diagram **NOT** accurately drawn

14 OABC is a parallelogram.

$$\overrightarrow{OA} = \begin{pmatrix} 1 \\ 2 \end{pmatrix}, \overrightarrow{OC} = \begin{pmatrix} 4 \\ 0 \end{pmatrix}$$

Diagram **NOT** accurately drawn

a Find the vector $\overrightarrow{OB}$ as a column vector.

X is the point on OB such that OX = kOB, where $0 < k < 1$

b Find, in terms of k, the vectors

i $\overrightarrow{OX}$ **ii** $\overrightarrow{AX}$ **iii** $\overrightarrow{XC}$

c Find the value of k for which $\overrightarrow{AX} = \overrightarrow{XC}$

d Use your answer to part **c** to show that the diagonals of the parallelogram OABC bisect one another.

4400 May 2006

15 A ball is dropped from a tower.

After t seconds, the ball has fallen a distance x metres.

x is directly proportional to t^2.

When $t = 2, x = 19.6$

a Find an equation connecting x and t.

b Find the value of x when $t = 3$

c Find how long the ball takes to fall 10 m.

4400 May 2006

16 A farmer wants to make a rectangular pen for keeping sheep.

He uses a wall, AB, for one side. For the other three sides, he uses 28 m of fencing.

He wants to make the area of the pen as large as possible.

The width of the pen is x metres.

The length parallel to the wall is $(28 - 2x)$ metres.

a The area of the pen is y m². Show that $y = 28x - 2x^2$.

b For $y = 28x - 2x^2$

i Find $\dfrac{dy}{dx}$

ii Find the value of x for which y is a maximum.

iii Explain how you know that this value gives a maximum.

c Find the largest possible area of the pen.

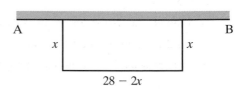

Diagram **NOT** accurately drawn

4400 May 2005

17 A fan is shaped as a sector of a circle, radius 12 cm, with angle 110° at the centre.

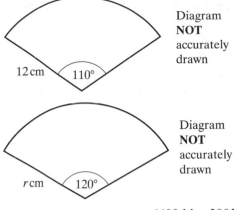

Diagram NOT accurately drawn

   a Calculate the area of the fan.

Another fan is shaped as a sector of a circle, radius r cm, with angle 120° at the centre.

Diagram NOT accurately drawn

   b Show that the total perimeter of this fan is $\frac{2}{3}r(3 + \pi)$ cm.

4400 May 2005

18 In order to start a course, Bae has to pass a test.
   He is allowed only two attempts to pass the test.
   The probability that Bae will pass the test at his first attempt is $\frac{2}{5}$
   If he fails at his first attempt, the probability that he will pass at his second attempt is $\frac{3}{4}$

   a Complete the probability tree diagram.

   First attempt   Second attempt

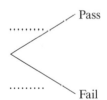

   Pass

   Fail

   b Calculate the probability that Bae will be allowed to start the course.

4400 May 2005

19 Solve the simultaneous equations
$$y = x - 1$$
$$2x^2 + y^2 = 2$$

20 A particle moves along a line.
   For $t \geqslant 1$, the distance of the particle from O at time t seconds is x metres, where
   $$x = 3t - t^2 + \frac{8}{t}$$

   a Find an expression for the velocity of the particle.

   b Find an expression for the acceleration of the particle.

   c Find at what time the acceleration is 0 m/s².

Paper 3

1 The bearing of town A from town B is 060°.
Find the bearing of town B from town A.

2 Rectangular tiles have a width x cm and height $(x + 7)$ cm.

$x + 7$

x

Diagram **NOT**
accurately drawn

Some of these tiles are used to form a shape.
The shape is 6 tiles wide and 4 tiles high.

width

height

Diagram **NOT**
accurately drawn

  a Write down expressions, in terms of x, for the width and height of this shape.**)**

  b The width and height of this shape are equal.

    i Write down an equation in x.

    ii Solve your equation to find an equation in x.

4400 Nov 2006

3 a Nikos drinks $\frac{2}{3}$ of a litre of orange juice each day.
How many litres does Nikos drink in 5 days?
Give your answer as a mixed number.

  b **i** Find the lowest common multiple of 4 and 6.

    ii Work out $3\frac{3}{4} + 2\frac{5}{6}$.
Give your answer as a mixed number.
You must show all your working.

4400 Nov 2006

4 The length of daisy stalks is given in the table below.

Length, x cm	Frequency
$0 \leqslant x < 2$	11
$2 \leqslant x < 6$	19
$6 \leqslant x < 8$	6
$8 \leqslant x < 12$	4

Calculate an estimate of the mean length of daisy stalk.

5 What is the effect of increasing £80 by 20%, then decreasing this value by 20%?

6 Mortar is made from cement, lime and sand.
The ratio of their weights is 2:1:9

Work out the weight of cement and the weight of sand in 60 kg of mortar.

4400 Specimen

7 Solve the inequality $7x - 3 \geqslant 4x + 15$.

8 Here is a four sided spinner.

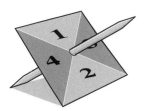

Its sides are labelled 1, 2, 3 and 4

The spinner is biased.
The probability that the spinner lands on each of the numbers 1, 2 and 3 is given in the table.

Number	Probability
1	0.25
2	0.25
3	0.1
4	

The spinner is spun once.

a Work out the probability that the spinner lands on 4.

b Work out the probability that the spinner lands on either 2 or 3.

4400 Nov 2005

9 *ABC* is a triangle
$AB = AC = 13$ cm.
$BC = 10$ cm.
M is the midpoint of *BC*.
Angle $AMC = 90°$.

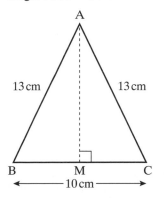

Diagram **NOT**
accurately drawn

a Work out the length of *AM*.

b A solid has five faces.

Four of the faces are triangles identical to triangle *ABC*.

The base of the solid is a square of side 10 cm.

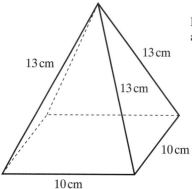

Diagram **NOT**
accurately drawn

13 cm

13 cm

13 cm

10 cm

10 cm

Calculate the total surface area of this solid.

4400 Nov 2005

10 a

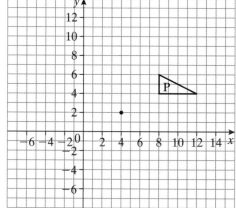

On a copy of the grid, rotate triangle **P** 90° anti-clockwise about the point (4, 2).

b

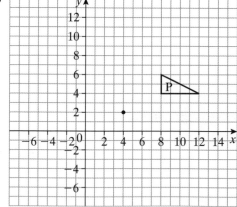

On a copy of the grid, enlarge triangle **P** with scale factor $\frac{1}{2}$ and centre (4, 2).

4400 Nov 2005

11 The table gives information about the ages, in years, of 100 aeroplanes.

Age (t years)	Frequency
$0 < t \leqslant 5$	41
$5 < t \leqslant 10$	26
$10 < t \leqslant 15$	20
$15 < t \leqslant 20$	10
$20 < t \leqslant 25$	3

a Copy and complete the cumulative frequency table.

Age (t years)	Cumulative frequency
$0 < t \leqslant 5$	
$0 < t \leqslant 10$	
$0 < t \leqslant 15$	
$0 < t \leqslant 20$	
$0 < t \leqslant 25$	

b On a copy of the grid, draw a cumulative frequency graph for your table.

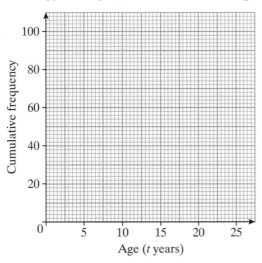

c Use your graph to find an estimate for the inter-quartile range of the ages.
Show your method clearly.

4400 Specimen

12 The straight line, **L**, passes through the points $(0, -1)$ and $(2, 3)$.

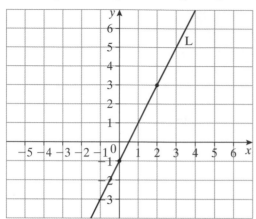

a Work out the gradient of **L**.

b Write down the equation of **L**.

c Write down the equation of another line that is parallel to **L**.

4400 Nov 2004

13 $p = 3^8$

a Express $p^{\frac{1}{2}}$ in the form 3^k, where k is an integer.

$q = 2^9 \times 5^{-6}$

b Express $q^{-\frac{1}{3}}$ in the form $2^m \times 5^n$, where m and n are integers.

4400 Specimen

14 a For the equation $y = 500x - 625x^2$, find $\dfrac{dy}{dx}$.

b Find the coordinates of the turning point on the graph of $y = 500x - 625x^2$.

c i State whether this turning point is a maximum or a minimum.
　　ii Give a reason for your answer.

d A publisher has set the price for a new book.
The profit £y, depends on the price of the book, £x, where

$$y = 5000x - 625x^2$$

　　i What price would you advise the publisher to set for the book?
　　ii Give a reason for your answer.

4400 Nov 2006

15 A solid hemisphere has a total surface area of $1000\,\text{cm}^2$ (including the base).
Find the hemisphere's

a radius　　　**b** volume

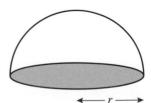

16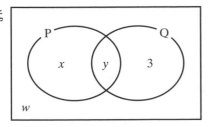

In the Venn diagram, 3, w, x and y represent the numbers of elements.

$n(\mathscr{E}) = 24$ $n(P') = 8$ $n((P \cap Q)') = 15$

a Find the value of   **i** w **ii** x **iii** y

b   **i** Find $n(P' \cap Q)$.

   ii Find $n(P' \cap Q')$.

   iii Find $n(P \cap Q \cap P')$.

4400 Nov 2005

17 The function f is defined as $f(x) = \dfrac{x}{x-1}$.

a Find the value of

  i $f(3)$,

  ii $f(-3)$.

b State which value(s) of x must be excluded from the domain of f.

c   **i** Find $ff(x)$.
    Give your answer in its simplest form.

  ii What does your answer to **c i** show about the function f?

4400 Nov 2006

18 Solve the simultaneous equations

$y = 2x - 7$
$x^2 + y^2 = 61$

4400 Specimen

19 Each student in a group plays at least one of hockey, tennis and football.

10 students play hockey only.
9 play football only.
13 play tennis only.
6 play hockey and football but not tennis.
7 play hockey and tennis
8 play football and tennis.
x play all three sports.

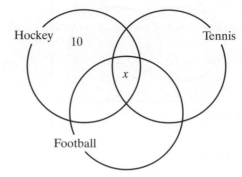

a Write down the expression, in terms of x, for the number of students who play hockey and tennis, but not football.

There are 50 students in the group.

b Find the value of x.

4400 Nov 2006

20 $\frac{1}{3}$ of the people in a club are men.

The number of men in the club is n.

a Write down an expression, in terms of n, for the number of people in the club.

Two of the people in the club are chosen at random.

The probability that both these people are men is $\frac{1}{10}$.

b Calculate the number of people in the club.

4400 Nov 2006

Paper 4

1 Work out the value in standard form to 2 significant figures of $\dfrac{3.23 \times 10^4}{\sqrt{1.81 \times 10^6}}$.

2 a Expand the following and simplify where appropriate.

    i $5(3x - 4)$

    ii $2x^2(3x - 4)$

    iii $(3x - 4)(3x + 4)$

    iv $(3x - 4)^2$

  b Factorise these.

    i $2xy - 12x^2y^3$

    ii $3x^2y - 9x^3y^2 + 15x^4y^3$

  c Simplify these.

    i $xy \times x^3y^5$

    ii $\dfrac{x^2y^7z^5}{xy^2z^3}$

3 Two fruit drinks, *Fruto* and *Tropico*, are sold in cartons.

  a *Fruto* contains only orange and mango.

    The ratio of orange to mango is 3:2

    A carton of *Fruto* contains a total volume of $250\,\text{cm}^3$.

    Find the volume of orange in a carton of *Fruto*.

  b *Tropico* contains only lemon, lime and grapefruit.

    The ratios of lemon to lime to grapefruit are 1:2:5.

    The volume of grapefruit in a carton of *Tropico* is $200\,\text{cm}^3$.

    Find the total volume of *Tropico* in a carton.

4400 Nov 2005

4 Here is a pattern of shapes made from centimetre squares.

  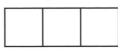

    Shape   Shape   Shape

   number 1   number 2   number 3

  This rule can be used to ffind the perimeter of a shape in this pattern.

> Add 1 to the Shape number and then multiply your answer by 2

  P cm is the perimeter of Shape number n.

  a Write down the formula for P in terms of n.

  b Make n the subject of the formula in part **a**.

4400 Nov 2006

5 Stella ran a 10 km road-race in 1 hour and 25 mins. Calculate her average speed in

  a km/h   **b** m/s.

6 $\mathscr{E} = \{$Integers$\}$
$A = \{1, 2, 3, 6\}$
$B = \{4, 5\}$
$C = \{x : 6 \leqslant 3x < 18\}$

   a List the elements of the set
      i $A \cup B$
      ii C

   b Find $A \cap B$.

4400 Specimen

7 Oil is stored in either small drums or large drums.
The shapes of the drums are mathematically similar.

Diagram **NOT** accurately drawn

A **small** drum has a volume of $0.006 \, \text{m}^3$ and a surface area of $0.2 \, \text{m}^2$.
The height of a **large** drum is 3 times the height of a small drum.

   a Calculate the volume of a large drum.

   b The cost of making a drum is \$1.20 for each m^2 of surface area.
      A company wants to store $3240 \, \text{m}^3$ of oil in large drums.
      Calculate the cost of making enough drums to store this oil.

4400 Nov 2004

8 $s = ut - \frac{1}{2}at^2$.
Find the value of S when $u = 10$, $t = 4$ and $a = -5$.

9 There are 48 beads in a bag.
Some of the beads are red and the rest of the beads are blue.
Shan is going to take a bead at random from the bag.
The probability that she will take a red bead is $\frac{3}{8}$.

   a Work out the number of red beads in the bag.
Shan adds some **red** beads to the 48 beads in the bag.
The probability that she will take a red bead is now $\frac{1}{2}$.

   b Work out the number red beads she adds.

4400 Nov 2006

10 Express 360 as the product of its prime factors.

11

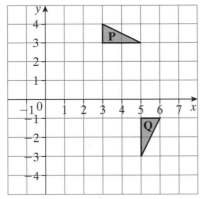

a Describe fully the single transformation that maps **P** onto **Q**.

b Another shape, **R**, is enlarged by scale factor 2 to give shape **S**.

Write down whether each of the following statements is a true statement or a false statement.

    i The lengths in **R** and **S** are the same.

    ii The angles in **R** and **S** are the same.

    iii Shapes **R** and **S** are similar.

4400 Nov 2005

12 Solve the following simultaneous equations

$$3x + 4y = 5$$
$$2x - 5y = 11$$

13 a Write the number 37 000 000 000 in standard form.

    b Write 7.5×10^{-5} as an ordinary number.

    c Calculate the value of $\dfrac{2.5 \times 10^{-3}}{1.25 \times 10^{7}}$ in standard form.

14 If $a = 3.7 \pm 0.5$, $b = 7.1 \pm 0.5$ and $c = 2.1 \pm 0.5$ and $y = \dfrac{a}{b - c}$, find to 2 significant figures

    a the maximum value of y.

    b the minimum value of y.

15

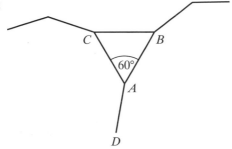

Diagram **NOT** accurately drawn

The sides of an equilateral triangle ABC and two **regular** polygons meet at the point A.

AB and AD are adjacent sides of a regular 10-sided polygon.

AC and AD are adjacent sides of a regular n-sided polygon.

Work out the value of n.

4400 Nov 2006

16 The sides of a triangle of perimeter 150 cm are in the ratio of 4:5:6.

Find the area of the triangle to 3 significant figures.

17

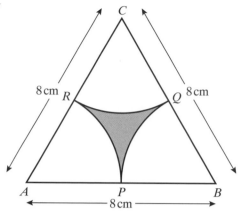

Diagram **NOT** accurately drawn

ABC is an equilateral triangle of side 8 cm.

With the vertices *A*, *B* and *C* as centres, arcs of radius 4 cm are drawn to cut the sides of the triangle at *P*, *Q* and *R*.

The shape formed by the arcs is shaded.

a Calculate the perimeter of the shaded shape.
Give your answer correct to 1 decimal place.

b Calculate the area of the shaded shape.
Give the answer correct to 1 decimal place.

4400 Nov 2004

18 a Complete the table of values for $y = x^2 - \frac{3}{x}$

x	0.5	1	1.5	2	3	4	5
y	−5.75	−2					24.4

b On a copy of the grid, draw the graph of
$y = x^2 - \frac{3}{x}$ for $0.5 \leqslant x \leqslant 5$

c Use your graph to find an estimate for a solution of the equation

$$x^2 - \frac{3}{x} = 0$$

d Draw a suitable straight line on your graph to find an estimate for a solution of the equation

$$x^2 - 2x - \frac{3}{x} = 0$$

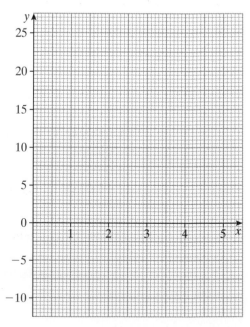

4400 Nov 2006

19 The unfinished table and histogram show information about the weights, in kg, of some babies.

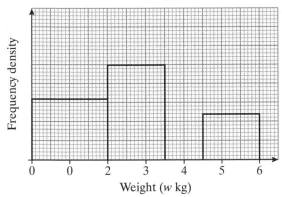

Weight (w kg)	Frequency
$0 < w \leqslant 2$	
$2 < w \leqslant 3.5$	150
$3.5 < w \leqslant 4.5$	136
$4.5 < w \leqslant 6$	

a Use the histogram to complete the table.

b Use the table to complete the histogram.

4400 Nov 2006

20

Diagram **NOT** accurately drawn

The diagram shows one disc with centre A and radius 4 cm and another disc with centre B and radius x cm.
The two discs fit exactly into a rectangular box 10 cm long and 9 cm wide.
The two discs touch at P.
APB is a straight line.

a Use Pythagoras' Theorem to show that $x^2 - 30x + 45 = 0$

b Find the value of x.
Give your value correct to 3 significant figures.

4400 Nov 2006